DEFENDING HUMANITY

When Force Is Justified and Why

George P. Fletcher
and
Jens David Ohlin

OXFORD
UNIVERSITY PRESS
2008

＃ 165478276

OXFORD
UNIVERSITY PRESS

Oxford University Press, Inc., publishes works that further
Oxford University's objective of excellence
in research, scholarship, and education.

Oxford   New York
Auckland   Cape Town   Dar es Salaam   Hong Kong   Karachi
Kuala Lumpur   Madrid   Melbourne   Mexico City   Nairobi
New Delhi   Shanghai   Taipei   Toronto

With offices in
Argentina   Austria   Brazil   Chile   Czech Republic   France   Greece
Guatemala   Hungary   Italy   Japan   Poland   Portugal   Singapore
South Korea   Switzerland   Thailand   Turkey   Ukraine   Vietnam

Published by Oxford University Press, Inc.
198 Madison Avenue, New York, New York 10016

www.oup.com

Library of Congress Cataloging-in-Publication Data
Fletcher, George P.
Defending humanity : when force is justified and why /
George P. Fletcher & Jens David Ohlin.
      p.   cm.
ISBN 978-0-19-518308-5
1. War.   2. War—Causes.   3. Intervention (International law).
4. Preemptive attack (Military science).   5. Security, International.
I. Ohlin, Jens David.   II. Title.
JZ6385.F54 2008
355.0201—dc22   2007033110

9 8 7 6 5 4 3 2 1

Printed in the United States of America
on acid-free paper

For Brachah
—G. P. F.

For Nancy
—J. D. O.

PREFACE

Since the U.S. invasion of Iraq in 2003, we have had good reason to worry about the future of the doctrine of international law prohibiting aggression except in cases of self-defense. It is not likely that an international court will prosecute members of the Bush administration for engaging in an aggressive war, and therefore it is critical that at least the literature of international law affirm the basic principles of mutual respect among nations. If those principles tolerate the use of force without Security Council authorization only in cases of self-defense, as Article 51 of the United Nations Charter tells us, then we need to probe the foundations of military self-defense more deeply than the literature of the past has done.

Two features of this book are novel. First, we draw on the analogy between self-defense in domestic law and in international law. An individual's being attacked at home or on the street bears a strong resemblance to a nation's being subject to foreign aggression. The difference is that we have centuries of legal experience analyzing self-defense in homicide cases and much less case law on the contours of international self-defense. To enrich the latter, it makes good sense to probe the theory of self-defense in domestic law—subject always to the condition that we recognize possible differences between individual and national aggression.

Second, to pursue this analogy we pay closer attention to the comparative law of self-defense than is usually done in conventional work on international law. It turns out that the French law of *légitime défense,* as well as other Continental analogues, offers an approach to self-defense much richer than we find in the English-language literature. Therefore it makes an enormous difference whether we follow the English or the French version of the United Nations Charter Article 51, which limits the use of force to situations called self-defense, or légitime défense.

This rich mixture of domestic and international legal theory is a product of our collaboration, a joint "criminal" enterprise born of our working together in various capacities at the Columbia Law School. I have undertaken very few projects with collaborators; largely I have found the ironing out of differences too time-consuming. With Jens, however, the common project proved to be both efficient and enlightening. I am indebted to him.

I wish to express my appreciation as well to Antonio Cassese, who seems to be among the first to understand the importance of interweaving domestic criminal law with the tradition of international law. Lenge Hong deserves thanks for her contribution to this project as well as for her work on several others in the past half-dozen years.

George P. Fletcher, New York, August 2007

Producing a book about war and self-defense in the current international climate has been a remarkable and rewarding challenge for us. The project began several years ago with a bold initiative: to bring a comparative analysis to bear on contemporary problems surrounding the use of force under international law. The book is comparative along at least two axes: the rules of self-defense in criminal and international law are analyzed comparatively, as are the relevant differences in domestic and foreign doctrines in criminal law.

Several times the book required revisions as new developments on the international scene demanded inclusion. In this still-evolving climate, the question of the use of force—its legality and its limits—is a problem that extends well beyond the legal academy. The question goes straight to the heart of the most crucial aspects of our public policy, aspects that all citizens must analyze as they exercise their democratic right to speak critically and vote.

Therefore, our theory of legitimate defense is of more than just academic concern. The practical consequences of our analysis can be seen in revitalized understandings of aggressive war, humanitarian intervention, and the Bush doctrine of preventive war, each of which we have devoted a full chapter to exploring. Humanitarian intervention deserves special mention because Article 51 places strict limits on the use of force, so scholars usually appeal to abstract norms of international law external to the Charter in order to defend the doctrine. Our theory of legitimate defense, however, provides a genuinely novel justification for humanitarian intervention through a new and refocused interpretation of Article 51 itself.

Throughout the writing process George proved to be an exciting intellectual collaborator, providing me with an inviting ear for new ideas and doctrinal innovation. It is rare to find a fellow scholar with a commensurate Weltanschauung—at least to such a degree that one can share that most precious of commodities: words. As we debated the ideas in this book, the only limits were reason and evidence.

Antonio Cassese provided a constant source of intelligence while we worked on this project, and we are grateful to him for creating a flourishing international academic community where questions of law and war are debated in New York, Florence, The Hague, and everywhere in between. Colleagues at Columbia Law School provided valuable feedback on our theory. Lori Damrosch and José Alvarez guided my understanding of international law from the very beginning, and Jonathan Bush taught me the history of war crimes law. Göran Sluiter of the University of Amsterdam provided insights on international criminal law. Carolijn Terwindt and Susanne Dyrchs composed the index. Jane Moriarty offered extensive comments on chapter 7 on preemptive force, while John Witt helped focus my thoughts on humanitarian intervention. I am personally indebted to Nancy Ohlin, who acted as a valuable sounding board for new and unconventional arguments.

Jens David Ohlin, New York, August 2007

CONTENTS

INTRODUCTION

In the past six years the United States has witnessed a series of events that have irrevocably changed how we think about war and self-defense. The attacks of 9/11 were followed by the invasion of Afghanistan and triggered a broader war on terror, overseen by the Bush administration through a variety of policy changes and legal justifications. Our enemies were taken from the battlefield, housed at Guantánamo Bay, and interrogated. They were labeled "unlawful combatants," a term invented by the U.S. Supreme Court during World War II and not used since then, but which the Bush administration pulled from the shelf and dusted off, effectively placing Guantánamo Bay in a legal black hole.[1] Then came the war in Iraq, the fall of Baghdad, an ongoing civil war there, and an impending nuclear threat from an increasingly militant Iranian government. Confronted with the possibility of more terrorist attacks, the CIA has engaged in extraordinary renditions, housed detainees in secret European prisons, and used torture as a strategic weapon against terrorism. Through all of this, the concepts of war and self-defense have been inextricably linked as we struggle to defend ourselves against our enemies and struggle with our own moral and legal doubts about the lengths and methods we may resort to. That we are entitled to defend ourselves is uncontroversial and cannot be questioned. But how we can do it is a separate question. The entire nation—president and citizens, academics and lawyers, legislators and voters—has struggled with these controversial issues.

None of these questions can be answered without a coherent understanding of the concept of self-defense and how it relates to war. Although the concept is well understood in the criminal law, its contours in the international sphere are only now being fully developed. Unfortunately, as so much has happened in the past five years, our legal doctrine has lagged behind. We are only now beginning to address the necessary legal questions.

Perhaps this is nothing new. During World War II, German saboteurs landed in a submarine off the coast of Long Island with a plan to disrupt U.S. military installations. They ditched their uniforms and blended in, caught by the FBI only after one of them had second thoughts and revealed the plot to the authorities. Roosevelt insisted that they be tried by military commission. They were convicted and sentenced to death, and the U.S. Supreme Court engaged in a hasty review of the constitutionality of the commissions before declining to step in.[2] The Court was under enormous pressure to decide quickly; the decision came just a day after the oral arguments, an unprecedented pace for a Court known for its careful deliberations. Six of the saboteurs were sent to the electric chair eight days later (the other two were spared by Roosevelt), and the Court finally released its written opinion more than two months after the fact. Even some current justices have questioned the value of the decision in *Ex parte Quirin*, calling it "not this court's finest hour."[3] Then, just as today, the law had difficulty catching up to the demands placed on it by the exigency of modern warfare.

This book aims to understand self-defense in war—its nature, its limits, its use as a justification. Chapter 1 discusses the need to reclaim the concept of self-defense and, in so doing, unite the fields of philosophy and law: two disciplines that have shown remarkably little interest in engaging one another on this issue. Chapter 2 surveys the landscape of current theories on self-defense by examining its roots in criminal law. Chapter 3 shifts the focus to the international context by examining the United Nations Charter and its provision on self-defense, Article 51, and argues for an extensive interpretation of self-defense that we call "legitimate defense." Chapter 4 then outlines the six essential elements of our doctrine of legitimate defense, and the remaining chapters apply the doctrine to the current controversies of the day. Consequently, chapter 5 discusses whether a nation can be excused for a wrongful invasion, and chapter 6 demonstrates how the doctrine of legitimate defense might justify a humanitarian intervention in another country. Chapter 7 tackles the thorny question of preventive and preemptive wars, a question of international law that has gained renewed urgency since the invasion of Iraq and the failure of the Bush administration to find weapons of mass destruction there; in that chapter we question whether the doctrine of legitimate defense could justify such an action. The final chapter examines the collective dimension of war, including the notion of collective guilt that each citizen bears for the actions of its nation.

This book is aimed at a broad audience. Of course, our interpretation of Article 51 of the United Nations Charter and our theory of self-defense will be of great interest to legal scholars and lawyers concerned with international law and the use of force. However, these questions of international law straddle the boundary between law and policy insofar as our legal argument also entails certain conclusions about what states may and may not do consistent with international law. Consequently, policymakers and politicians should be especially concerned with our argument, because it is they who must decide when and where to send our military forces. The law of war cannot be restricted to legal scholars who rework the doctrine in the privacy of the academy. International law constrains the behavior of states, and therefore we direct our argument to our political leaders. If Article 51 and the law of international self-defense is to be reinterpreted, surely it is they who must hear it.

Finally, and most important, our argument is aimed at the general public. International law has now emerged as a central element of our political discourse, not just with regard to the war in Iraq but whenever military conflict is considered. This represents a shift in our dialogue from a generation ago during the Vietnam War. This is not to suggest that international law was entirely absent from our nation's political discussions during the Vietnam era. Rather, it simply recognizes that international law has penetrated deep into the four corners of our political landscape. Also, this is not to say that everyone, especially in government, is appropriately schooled in the subtleties of international law. Indeed, the Bush administration has demonstrated a marked hostility to international law by explicitly denying that their actions should be constrained by international law. But while this attitude may exist in some corners of the current State Department, it is not representative of the American public. When we, as a nation, debate military action in Iraq, Afghanistan, or elsewhere, the American public asks, with great sophistication, whether the military action would be lawful under international law. The discussions are not just confined to our sense of morality. The greater American public is now increasingly cognizant of our national obligations under international law and how this might impact the decisions that all of us, as a democratic nation, participate in. Just as Michael Walzer brought the idea of a just war into political currency a generation ago, now the notion of a legal war is being brought into political currency, and so these crucial questions of international law must be debated by both legal scholars and the public alike.

Although there are many challenging and provocative arguments in this book, they can be understood by everyone. No prior knowledge is required. One need not be a scholar of international law, nor even a lawyer, to participate in this discussion. Although we have included notes to document our extensive research and analysis, general readers need not be concerned with the notes unless they are interested in delving further into a particular area. The notes are also provided for scholars who wish to continue the debate on a particular point of law or philosophy, an outcome we would certainly welcome. The many arguments in these pages are not the end—they are just the beginning—as we enter a new period of relevance and urgency for international law and the law of war. All are invited to join in.

DEFENDING HUMANITY

1

MURDER AMONG NATIONS

This book takes as its starting point the unquestioned truth that military action is justified under international law only in two situations. First, the United Nations Security Council can authorize the use of military force to restore collective peace and security. Second, military action is legal even in the absence of Security Council authorization, but the United Nations Charter explicitly limits this use of military force to self-defense, and only in cases in which an "armed attack" has occurred. Because Security Council authorizations for military force are rare, the universally accepted formula for justifying aggression is the combination of two magic words: *self* and *defense*. If your sense of self is big enough, however, you might invoke the principle of self-defense any time your interests are threatened. It is not surprising, then, that dictators and tyrants have also invoked this magic formula. Osama bin Laden and his followers consider the presence of American bases in Saudi Arabia a form of aggression, an intrusion of Western elements in Dar el Islam; thus they bomb symbols of Western power in New York and Washington to defend "the Home of Islam."

The rhetoric of self-defense was one of the most frequently abused concepts of the twentieth century. As Oscar Schachter wrote in 1989, "When they [states] have used force, they have nearly always claimed self-defense as their legal justification."[1] This is true in internal politics as well. Stalin claimed he was defending the Soviet Union by engaging in a campaign of homicidal elimination against the Kulaks, bourgeois farmers who supposedly stood in the way of collectivized agriculture. Hitler used an image of Jews as insects attacking the integrity of the German *Volk* in some of his most graphic anti-Semitic propaganda.[2]

In the United States, too, politicians have fallen back on self-defense as the metaphor of first resort. They believed that the first major threat after World War II was the Communists. Many people feared that the United States would go the way of Eastern Europe—that we, too, would suffer knocks on the door in the middle of the night. The House Un-American Activities Committee was supposedly the vanguard of defensive operations: committee members would find out who the disloyal (or potentially disloyal) were, and then it was up to the rest of American society to shun them in order to keep our institutions free of Communist contamination. Senator Joseph McCarthy sought to expose one Communist after another, often on the basis of rumor and association. His campaign collapsed when he accused the U.S. Army itself of being a danger to our internal security. The lawyer Joseph Welch stood him down with a single question: "At long last, sir, have you left no sense of decency?"

Welch engaged in verbal self-defense—an often overlooked variety—and used a deft phrase to put McCarthy in his place. The country held firm against those who wanted to use the fear of Communism to undermine the legal rights of ordinary citizens. When Congress wanted to make it a crime to be a mere member of the Communist Party, the courts said no, membership alone is not a form of aggression and does not warrant the use of criminal sanctions in self-defense.[3] Criminals must pose an active threat, something more serious than signing up and getting a membership card.

The next threat to American security was the wave of criminal behavior that hit the streets in the 1970s and 1980s, a threat that made residents of the racially mixed areas we dubbed "inner cities" quiver in their shoes as they walked to the market. In New York City, a Columbia University law professor was murdered in broad daylight in 1972, six blocks from the law school, in a neighborhood where being white was seen as a provocation.[4] In 1984, as the violence approached its peak, a black teenager traveling with three rowdy friends approached Bernhard Goetz on the New York subway and asked him for five dollars. Goetz responded by pulling out a concealed .38 Smith & Wesson revolver and shooting all four "threatening" youths, one in the back and another allegedly while sitting down. No one died, but the last teenager hit, Darrell Cabey, was paralyzed for life. Goetz's trial proceeded amid widespread fear of the perceived enemy on the streets and corresponding sympathy for the new "hero" of New York: the subway vigilante who stood up against the "terrorists" of the time.[5]

In those days we heard the recurrent conservative cry that we were too soft on the real criminals (not Goetz, but the four black kids who got shot) and therefore we must restrict civil liberties to ensure our safety on the streets. Restricting civil liberties was repeatedly advocated in the name of self-defense. This was called "balancing liberty against security." As David Cole insightfully pointed out, we balanced "their" liberty against "our" security.[6] This is the essence of defensive action: we harm others for our own sake.

Self-defense is always directed against an outside aggressor, and those who are alien, foreign, or different are far more likely to be seen as the "other." Over the past half-century our image of danger has continually shifted: Communists were the original postwar other, then came the street criminals, particularly the black criminals typified by the four "punks" in the *Goetz* case. Since 9/11, the other has been terrorists, in particular Islamic terrorists. Claims of self-defense are always easier to level against the foreign, the different, or the despised.

1. RESCUING SELF-DEFENSE

Self-defense is therefore the ubiquitous, all-purpose justificatory argument. Virtually every wrongdoer appeals to self-defense, no matter how aggressive he or she might be. But if everything is defensible, then nothing is defensible. All-purpose concepts lose their traction and slide into trivia and tautology. This has happened lately to claims of self-defense in international politics. The modern world is rife with terrorism, military aggression, and preemptive war, but no matter who does the killing, the state always appeals to the rhetoric of self-defense. Al Qaeda claims to be defending itself against American imperialism; the United States defends itself against Islamic terrorists. The Palestinians defend themselves against Israeli aggression; Israel defends itself against Palestinian terrorism. The rhetoric of self-defense is so appealing that no one is willing to surrender it. Ever more reason, then, that the confusing strands of self-defense need to be untangled. We still need the distinction between aggression and self-defense too much to surrender it to those who would abuse it. We must defend the notion of self-defense itself.

The need to rescue the concept is most acute at the level of international law, where we encounter so many allegations—but few authoritative judgments—of right and wrong. The few examples we do have, such as the *Nicaragua* case before the International Court of Justice, and multiple cases involving Israel (against Egypt in 1967, against Iraq in 1981, and, recently, involving the building of a security wall on the West Bank) are mined for all they are worth.[7] But these few examples are not enough to serve as the foundations of a nuanced body of law. As it stands today, international law is still too thin to provide a reliable framework for understanding even such basic concepts as aggression and defense.

The solution is to draw on the leading domestic legal systems, which have a very rich jurisprudence of statutes and decisions on the permissibility of self-defense. As self-defense was the central issue in the *Goetz* case, it is also the anchor that orients the law of war. As four black youths surrounded and threatened Bernhard Goetz on the subway, at least four Arab states aimed their hostile power against Israel in June 1967. If Goetz could use necessary force to prevent being mugged, then our tentative judgment should be that Israel could also use necessary force to prevent being overrun.

But recent events have revealed a troubled understanding of self-defense in international law, not just among diplomats but among legal scholars and commentators as well. The first event was the Bush administration's self-declared War on Terror. After the attacks on 9/11, the Bush administration launched a comprehensive program at home and abroad—all in the name of self-defense. The first targets were the Taliban and Al Qaeda forces in Afghanistan. But the Bush administration soon appealed to the right of self-defense for a series of more aggressive moves: military detentions at Guantánamo Bay, increased domestic police powers under the Patriot Act, the right to declare American citizens unlawful combatants and detain them without charge, military commissions with few procedural protections for unlawful combatants, extraordinary renditions of terror suspects to friendly countries for their interrogation and torture, and the official sanctioning of torture at secret CIA detention facilities in Europe. Although all of these actions raised numerous legal problems under international and U.S. constitutional law, all of them were explicitly justified by our right to self-defense against terrorism.

The second event was the invasion of Iraq. Not content with military strikes in Afghanistan, the Bush administration launched the invasion of Iraq after arguing to the American public, the United Nations, and the world that Saddam Hussein had weapons of mass destruction that constituted an unacceptable threat against U.S. security. Preemptive war against Iraq was therefore justified on the grounds of our self-defense. At the same time, several U.S. intellectuals justified the invasion on the grounds that the United States should come to the aid of oppressed minority groups in Iraq that had long since suffered under Hussein's oppressive regime. This was an appeal to "the defense of others," or the idea that all individuals have the right to come to the aid of those who have been unjustly attacked. Of course, the weapons of mass destruction were never found, prompting questions about manipulated intelligence and the degree to which a preemptive war of self-defense can be justified on the basis of false information.

The third event was the Israeli military incursion into Lebanon in the summer of 2006 to stop Hezbollah guerrillas from launching rockets across the border. After the unilateral withdrawal of Israeli forces from southern Lebanon in 2000, Hezbollah guerrillas took up positions there and began firing rockets across the border into Israeli territory. These Hezbollah attacks intensified in the summer of 2006, with frequent rocket launches and civilian casualties. In most cases the rockets were crudely targeted or not targeted at all, and Hezbollah guerrillas simply lobbed the rockets indiscriminately at an Israeli town with no particular target—civilian or military—in mind. Because the Lebanese government was either unwilling or unable to stop the attacks, Israel appealed to its right of self-defense and launched a ground offensive into southern Lebanon to neutralize the guerrillas. This campaign was both militarily and politically disastrous for the Israelis: they had difficulty neutralizing the far-outnumbered guerrillas and convincing the world that their actions were legally justified.

At the same time, a genocidal civil war rages in the Darfur region of Sudan. Government forces are engaged in a bloody war against rebels and have armed sympathetic militia known as Janjaweed to conduct military operations on their behalf, many of which have resulted in civilian casualties, massacres, rape, and forced displacement. The UN Security Council referred the situation to the International Criminal Court, whose prosecutor is considering charges of war crimes, crimes against humanity, and genocide. The government of Sudan has attempted to justify these widespread attacks against civilians as unfortunate collateral damage from their anti-insurgent campaign against the rebels. In other words, the government is appealing to its right of self-defense as a legal defense for war crimes.

These four ugly situations have brought the right of self-defense into the very forefront of our political and legal discourse, although in each case there is profound misunderstanding about the structure of this right, its scope and limits, and its application in the moral and legal arenas. Although the right of self-defense has served as a bedrock legal principle in international law since the adoption of the UN Charter in 1945, it remains somewhat underanalyzed, especially when compared to the principle of self-defense in domestic criminal law, where a well-considered doctrine and rigorous theory have flourished. The disparity might stem from the fact that disputes about self-defense in international law are rarely decided in a legal forum, whereas disputes about self-defense in criminal law must always result in a decision one way or the other that must be justified by a judge in writing, either in jury instructions or in a written opinion, thereby enriching the materials of the law. Although some disputes are brought before the International Court of Justice and legal scholars have produced commentaries on the use of military force under international law, there are basically few forums for actual legal adjudication, especially when one considers that claims of self-defense are made in almost every international military dispute. The result is a comparative advantage for the domestic law of self-defense over the international one.

Of course there are many hotly contested issues in the domestic law of self-defense, too, and the same issues, or most of them, appear in these international law cases. By studying the debates in domestic law, we will uncover sources of knowledge and understanding that, by and large, have been ignored in the literature of international law—as though there were a hidden manuscript containing the secrets of life but no one had bothered to read it.

2. THE COLLAPSE OF INTERNATIONAL LAW WITH CRIMINAL LAW

Part of the confusion stems from a basic misunderstanding about the relative roles of international law and international *criminal* law in contemporary discussions of the use of force. This is especially the case during discussions about proportionality, a principle that we focus on in more detail in chapters 4 and 8. For example, during the Israeli campaign against Hezbollah, legal commentators often worried that the military response was not proportional, although few were clear about whether they meant that Israel's *actions* were disproportionate because its military objectives went too far, or whether Israel's *methods* were disproportionate because they caused too many civilian casualties. Although both questions are about proportionality, the former is about international law and the latter is about international criminal law, and the principle of proportionality works somewhat differently in each area. The word is the same, and to a certain degree the concept is the same, but the questions are different.

To understand this distinction, we must be clear about the contours of international law and criminal law. This is elementary, of course, although it is shocking how many international and criminal lawyers confuse these basic ideas. International law governs relations between nation-states who voluntarily consent to be bound by treaties through which they gain international rights and responsibilities with regard to other nation-states. The UN Charter, the Security Council, and customary international law may also impose duties on nation-states that are not voluntary, but they are still state responsibilities toward other states. For example, states are restricted in their use of military force, as explained earlier, whether they have signed a treaty to that effect or not. If they use military force without Security Council authorization, it must be in self-defense, and, presumably, the response must be proportional. They may use as much force as is necessary to protect their political independence and territorial integrity. They cannot legitimately continue a war of self-defense as a quest for vengeance.

Some treaties, such as the Geneva Conventions, impose restrictions on how nation-states can treat individuals, either their own nationals or nationals from other countries who, for example, they capture on the battlefield. Although in this case the beneficiaries of these treaties are individuals, the basic international obligation is owed by one nation-state to another. It is a promise, as it were, that each nation makes to the international community of nation-states to respect certain human rights. Violations of these human rights treaties yield state responsibility. One state may call another state to account for its horrendous treatment of individuals during war or peacetime. Violations of international law can be enforced by the Security Council, by economic sanctions, by actions before the International Court of Justice, or by public disapprobation before the court of world opinion.

International criminal law, on the other hand, imposes criminal liability on individuals for a very specific set of crimes that have been identified by international criminal statutes, such as the Rome Statute. These statutes typically define acceptable behavior during war, whether international or internal, and they outlaw genocide, war crimes, crimes against humanity, and torture. Individual soldiers and militias can be found guilty of these offenses, as well as their military and civilian supervisors who may have ordered them to commit these crimes or negligently supervised them and allowed the crimes to happen under their watch. Either way, it is individuals who will be prosecuted before an ad hoc tribunal such as the International Criminal Tribunal for the Former Yugoslavia or the International Criminal Tribunal for Rwanda, or the now permanent International Criminal Court in The Hague, and sentenced to prison. Simply violating an international human rights treaty is insufficient here. Prosecutors must prove that the defendants engaged in particular conduct that meets the definition of an offense that is recognized in international criminal law as one that bears *individual* criminal liability.

To take one example of how difficult these matters can be, consider that the Nazis were charged at Nuremberg with conspiracy to wage a war of aggression. While it was undoubtedly true that as a political matter the Nazis had engaged in aggressive war by invading Poland, marching across Europe, and invading Russia, the legal case was more complicated. The prosecution, led by a team of U.S. and international lawyers from Allied countries, appealed to the Kellogg-Briand Pact, an international treaty signed in the aftermath of World War I in which the signatories, including Germany, pledged to renounce aggressive war.[8] This treaty was introduced to show that aggressive war was a violation of international law. But this posed a problem. The prosecution needed to show that aggressive war was not only a violation of international law, implying state responsibility for Germany as a whole, but that it was also a crime bearing *individual* criminal liability for those who conspired to wage aggressive war. This point was finessed quite delicately by the prosecution; the judges of the International Military Tribunal held that the Nazis could not claim ignorance that aggressive war was wrong.

Similar confusion about international law and international criminal law reigned during the Israeli war with Hezbollah. It is imperative that we separate two distinct questions. The first is how far Israel can go on behalf of its right of self-defense: Does self-defense allow Israel to bomb Hezbollah targets, to bomb government targets in Beirut, to invade southern Lebanon, to invade the whole country and topple the government? What is the proportional response? These are all questions governed by standards of international law for the exercise of self-defense. The second question is how many civilian casualties are acceptable when an individual military commander authorizes an attack against a specific target. According to the Rome Statute, the attack could be a war crime if the civilian casualties are disproportionate to the importance of the military target. As we shall see in chapter 8, determining what is meant by "disproportionate" in these circumstances is a difficult affair. But the important point at this juncture is that proportionality here is judged at the level of the individual target and the collateral damage that could result. A disproportionate attack—whatever that is—would violate international criminal law, and soldiers could go to prison for it. This really has little to do with whether Israel was justified in invading southern Lebanon. One has to do with justifying a decision to go to war, the other with how the war is conducted. As we shall see, this is a familiar distinction in discussions about just war theory, though lawyers do not always pay sufficient attention to it. It is possible to conduct a just war unjustly, just as it is possible to conduct an unjust war justly.

These difficulties stem from our hesitation over whether we want to hold nations or individuals responsible when war goes wrong. The first is a function of international law and the second a function of international criminal law; this tracks a basic distinction between collective and individual conduct. When nations act *as a collective group*, certain standards apply, and we hold states responsible when they fail to live up to them. When soldiers act *as individuals*, criminal law applies, and we put them on trial when they break the law. There is much to learn by comparing self-defense in these two arenas. This dialectic runs throughout the entire book.

3.   A WORD ABOUT METHODOLOGY

We must be clear about the structure of our reasoning and the mode of analysis we employ. Collective and individual actions are different, but the concepts underlying each are the same. We just finished explaining the differences between criminal and international law. So what are the similarities between criminal and international law, between the actions of individuals and the actions of nation-states? Both are free agents, capable of deliberating and making choices and can be held responsible for their actions.[9] Whether or not a nation is actually a single moral agent—as opposed to just a jumbled collection of individual human beings—is beside the point. This worry is irrelevant because international law *treats* nations as moral and rational agents, regardless of the degree to which they actually approximate this ideal. International law grants legal *personality* to nation-states and treats them as the kind of entities that have moral and legal interests, rights, and responsibilities.[10] They can be held accountable for their actions just like human beings.

The legal personality of nation-states is what allows us to compare self-defense in international law with self-defense in domestic criminal law.

It is therefore possible to import some of the basic intuitions of criminal law and apply them to the sphere of international relations. We can take well-traveled distinctions from the criminal law of self-defense and bring these tools to bear on the conundrums of self-defense that we find at the international level. However, we cannot simply import basic notions of domestic criminal law blindly. This would be foolish. We do not regard criminal law as providing an intellectual foundation for international law. Criminal law is not primary as a matter of logic, nor is it necessarily more developed in every case. It does not yield essential truths that should be accepted without question. Our methodology is not simply to apply criminal law notions to the field of international law. That would be naïve.

Rather, we must critically examine whether the principles of criminal law generate just outcomes in the sphere of international law. To the extent that these doctrinal devices from criminal law improve our understanding of international relations, we should consider accepting them as amendments to the basic principles of international law. For example, if we have solid criteria for what kind of attacks spark a right to self-defense in criminal law, we should take these criteria, apply them to nation-states, and see what the result might be. The result might be a deeper and more sophisticated doctrine of the international law of self-defense. If, on the other hand, we analyze a basic principle of criminal law and find that, when applied to the international arena, it yields absurd consequences, it would be unwise to import it into international law. Indeed, we might even revisit the principle's application in criminal law if we see that it has strange consequences at the level of collective action. This would be a warrant for serious intellectual reconsideration for the criminal lawyer as well.

So the process works both ways. We can also take general principles of international law and see how they would fare in the world of criminal law. The best procedure is a rich interplay between the two spheres, charting the consequences of basic principles and making revisions to basic elements of both criminal and international law as warranted. Neither should be regarded as primary, free from skepticism, and immune from revision. John Rawls famously described this procedure as one of "reflective equilibrium": a coherent balance between general principles and individual judgments that are evaluated in relation to each other.[11] For example, in his famous *Theory of Justice*, Rawls described the basic principles of justice by formulating a thought experiment. What principles of justice would we agree to in an original position if we had to bargain behind a veil of ignorance that prevented us from knowing anything about our lives? We would have no information about our race, sex, ethnicity, talents and handicaps, or our economic class in the original position. Bargainers at the social contract table would be bare rational agents, deciding on pure reason alone on principles that, by definition, would be fair for everyone. We would agree on principles fair to everyone because each of us has no idea who we really are.[12] The thought experiment was therefore constructed to write out self-interest and replace it with pure, disinterested, rational deliberation. One might think Rawls's thought experiment involved *deriving* the basic principles of justice from the structure of this original position. But this would be a false interpretation. Rather, the process is one where we develop our intuitions about what would be a fair structure for the original position. For example, it is clearly fair that we stand behind a veil of ignorance. Otherwise, we would just choose principles of justice that benefit only ourselves, and this could never yield adequate principles of justice. With a basic procedure in place for the negotiations, we can see which principles of justice these negotiations would yield and therefore evaluate our intuitions at this end of the spectrum as well. If the principles of justice generated by this procedure are morally suspect (i.e., they strike us as wrong), we go back and revise the basic structure of the original position. And the process works in reverse, too.

Legal scholars are sometimes unused to this style of reflective equilibrium because the law is more hospitable to top-down thinking. We take a constitutional provision as fixed and foundational and determine whether a particular statute transgressed it; we consider a penal law as fixed and we determine whether someone's conduct violated it. But as many legal scholars have recognized, one must be careful not to descend into formalistic thinking, of blindly applying principles to new situations without also reevaluating a principle's application to the new context.

In our case, one reason to engage in this process of reflective equilibrium is a belief that both international and criminal law share, at least in part, a deep structure that unifies the two fields. This is not to say that there is some unified body of legal principles from which all of criminal and international law could be derived. Nothing is further from the truth. Rather, the hypothesis is simply that both fields regulate the actions of morally free agents in some way. Specifically, both fields regulate when a free agent may use force in self-defense. It is reasonable to expect that there is enough of a common deep structure here to make a process of reflective equilibrium worthwhile. And it is quite clear that international lawyers are all too ignorant of this deep structure and its application in criminal law, and criminal lawyers are disengaged from how this deep structure plays out in the field of international relations. It need not be so.

It goes without saying that these types of comparisons are dangerous and fraught with difficulty. It is quite possible to gloss over the relevant differences and find similarities where there are none. We might, as it were, *exaggerate* the common structure between the two areas, leading to an impoverished understanding of the law and urging revisions to international law that are supported by nothing more than bare intuition and gut feelings. The danger is accentuated by the fact that this comparison between the international law of self-defense and the criminal law of self-defense has never been accomplished before, or at least has never been given anything other than a cursory treatment. But the difficulties inherent in a comparative analysis are no reason to preempt a legal analysis before it even starts. They are only a reason to pledge fidelity to *both* international and criminal law and to avoid parodying either of them while seeking a deeper understanding of both.

4. LAWYERS AND PHILOSOPHERS

Our methodology involves more than just a comparison between international and criminal law. Two other modes of inquiry are also synthesized by our work: law and philosophy. The U.S. wars of the past thirty-five years have triggered an intense reaction not only among lawyers but also among philosophers, political theorists, and legislators. Each of these branches of knowledge has a different way of thinking and a different take on war, its ethics, and its principles of legitimacy. Of particular concern to us is the influence that training and institutional loyalties have on the way lawyers and philosophers think and write about war. The remarkable fact is that these two camps ignore each other. Take the influential philosophical text by Michael Walzer and the leading legal text on war and self-defense by Yoram Dinstein. Both books are currently enjoying third editions published in the twenty-first century, yet neither takes notice of the other.[13]

The international criminal lawyers who write about the Rome Statute are equally indifferent to philosophical and jurisprudential analysis. The Rome Statute rather clearly reflects the principle of "double effect" devised by Aquinas to reconcile self-defense with respect for human life.[14] Originally, the principle maintained that killing an attacker was permissible (despite the usual prohibitions against killing) because one's use of force had the primary effect of self-protection: the killing of one's attacker was simply a secondary or "double" effect that could not be avoided. Indeed, the killing was not even intended, in a sense, because one's primary and direct intention was simply to save oneself. When applied in military contexts, the principle holds that it is legitimate to bomb military targets provided that the secondary effects to the civilian population are not disproportionately onerous. Thus, as the argument goes, there is a radical difference between targeting civilians and causing one hundred deaths (as a primary effect) and aiming at a military installation with the "collateral effect" of one hundred civilian deaths. The Rome Statute incorporates this principle by penalizing a number of offenses based on intentionally launching attacks on civilians, including one undertaken in the knowledge that such attack will cause incidental loss of life or injury to civilians or damage to civilian objects or widespread, long-term, and severe damage to the natural environment that would be clearly excessive in relation to the concrete and direct overall military advantage anticipated.[15]

In other words, directly attacking civilians is illegitimate, as is aiming at a military target if such a strike anticipates excessive civilian harm relative to the expected military advantage. Yet it is important to distinguish this standard of "excessive" collateral harm from a straightforward utilitarian analysis based on the expected costs and benefits of the action. The military purpose permits a possible skewing of the balancing process in favor of the military. This analysis follows from the principle of double effect; many moral philosophers explicitly subscribe to this doctrine or a version of it,[16] yet the literature of international law appears to be indifferent to the philosophical roots of the Rome Statute. This degree of specialization is unfortunate. It clearly renders the legal commentary less convincing and deprives the law of war of its proper cultural resonance.

In this divide between philosophy and law, neither side concedes that there may be anything to learn from the other.[17] To make the positions more nuanced, a third approach, exemplified by Walzer's important book, stresses the evolution of moral arguments in their historical context. Though it was the specific issues of the Vietnam War that galvanized a generation of moral philosophers to take up the law of war,[18] their mode of argument is timeless. They emphasize moral reasoning, abstracted from concrete historical examples. For example, Thomas Nagel argues for an "absolutist" position based on the Kantian view that we must respect others as persons rather than things. In his view, the utilitarian approach to war reduces the human beings on the other side to things.[19] His test for treating someone as a person is whether you could plausibly explain what you are doing to him or her and expect the person to understand.[20] You can expect soldiers to understand why you are shooting at them, but you cannot expect civilians to show a similar sympathy for their being the object of aerial bombardment.[21]

We find ourselves caught between these camps, loyal to all, fully committed to none. We also reject the need to make a choice between them. Any serious book about international law and justice must pay attention to history and philosophy as well as law. Philosophers who ignore the law do so at their peril. The lawyers who ignore moral and analytic philosophy—not to mention great European thinkers like Karl Jaspers— isolate themselves from the traditions of Western humanistic thought. Historical sensitivity requires attention not only to specific cases, but also to the evolution of the relevant distinctions in the law of war, among them the distinction between *jus ad bellum* (just war) and *jus in bello* (justice in the conduct of war).

In the hope of overcoming some of these professional divisions, we will specifically explore the differences between philosophers and lawyers when they write about the law of war. The three most salient dimensions of comparison are the role of authority, the relevance of institutions, and the problem of individualism and the nation as an actor in international affairs. Discussing these points of difference between philosophers and lawyers provides an opportunity to introduce many of the themes of this book.

A.   *The Role of Authority*

Lawyers begin their arguments by relying on authoritative texts: statutes, cases, and regulations. The relevance of authority means that an *avocat* arguing under French law will make different arguments from a lawyer invoking common law sources. Some lawyers put a great deal of weight on commentators; others regard commentators as secondary sources.[22]

International lawyers appeal to a shared set of authoritative sources, among them the Hague and Geneva Conventions. Philosophers, when they reason from first principles, can reach some of the same results found in these conventions. For example, Nagel offers an account of the ban on dumdum bullets and other weapons that inflict more harm than is necessary to achieve their military objectives.[23] Similarly, the philosopher Richard Brandt offers a rule-utilitarian account for many of the same rules.[24] Reasoning from first moral principles, however, philosophers cannot account for many of the most entrenched rules in the law of war, such as the bans against treachery and the use of poison. We need to examine the history of these prohibitions and then offer an account of why standard approaches to moral philosophy fail to account for them.

For example, Article 23 of the Hague Convention of 1907 addresses poison and treachery side by side in subsections (a) and (b):

> Art. 23: In addition to the prohibitions provided by special Conventions, it is especially forbidden
>   (a) To employ poison or poisoned weapons.
>   (b) To kill or wound treacherously individuals belonging to the hostile nation or army.

These stable and venerable provisions in the law of war are now incorporated word for word into the Rome Statute.[25] But why are these activities prohibited, and do the two provisions have anything to do with each other?

Moral philosophy has little to offer us by way of explanation. Whether you agree with Nagel that soldiers must treat the enemy as persons, or you side with the utilitarians, the relevant moral arguments focus on the victims, their interests, and their rights. Either the focus is on the status of others as persons, or on their interests in the utilitarian calculus. Neither approach can accommodate an ethic of honor.

The only way to understand the ban against poison and treachery is to invoke the chivalric tradition of warfare. The honorable way to fight is to face the enemy and kill by shooting, dueling, stabbing—even strangling. Poison is dishonorable because it is secret and duplicitous. Treachery is prohibited for similar reasons. Approaching the enemy as a civilian and then opening fire or pretending to be wounded before attacking exploits the law of war for unfair advantage. In warfare it is permissible to stand and deliver—to look the enemy squarely in the eye and shoot; it is not permissible to pretend to be innocent and then use poison to contaminate his food, his land, or his wells.

The association between poison and treachery derives, to be sure, from the chivalric tradition of fighting as a knight and a gentleman. But perhaps the best philosophical account is found in Immanuel Kant's treatment of jus in bello:

> Means of defense that are not permitted include using its own subjects as spies; using them or even foreigners as assassins or

poisoners [emphasis added] (among whom so-called snipers, who
lie in wait to ambush individuals, might well be classed); or
using them merely for spreading false reports—in a word, using
such underhanded means as would destroy the trust requisite to
establishing a lasting peace in the future.[26]

This strong taboo against deception lies deeply entrenched in Kantian
thinking. To the dismay of many modern readers, Kant also treats lying
as an absolute evil—with no possibility of justification—but his transition
from the morality of deception to a prohibition against spies and snipers
is admittedly dubious. (Of course, spying is not prohibited today under
the law of war. On the contrary, the Hague Convention of 1907 seeks to
protect spies against immediate execution, and snipers are often considered heroes of modern warfare.)[27] The important point is that Kant con-
nects the issue of poisoning with the kind of dishonorable behavior that
bars future relations of mutual respect.

Kant provided the primary source for jurist Francis Lieber's code of
1863, which makes the direct association between poison and perfidy in
Article 16:

Military necessity does not admit of...poison in any way, nor of
the wanton devastation of a district. It admits of deception, but
disclaims acts of perfidy; and, in general, military necessity does
not include any act of hostility which makes the return to peace
unnecessarily difficult.

"Perfidy" and "treachery" are synonyms and are used interchangeably. If
men fight dishonorably, they cannot return, with self-respect, to a culture
of peace; therefore, as Lieber argues, perfidy is banned because it "makes
the return to peace unnecessarily difficult." If the state is at war and its
soldiers engage in forbidden means of warfare—poison and perfidy—it
humiliates and demeans the adversary and makes it more difficult for the
two sides to return to peaceful postwar relations.

To sum up this line of thought, poison is treachery because it kills
secretly; it works by laying a trap for the unwary. Even a seemingly innocent meal can be an occasion for treachery. It is widely believed that the
Ukrainian leader Victor Yushchenko suffered dioxin poisoning because
he happened to have dinner with the wrong person. Yushchenko was in
the midst of a tight and controversial election battle for the Ukrainian
presidency in September 2004 when he became severely ill. Images of him

in the hospital with a severely disfigured face—one symptom of dioxin poisoning—were flashed across television broadcasts. Yushchenko lost the first round of voting by less than 1 percent, but allegations of electoral fraud and massive street protests led to a new election, which he narrowly won. The poisoning prompted charges of covert agency involvement, although the case was never solved by the authorities. Whoever poisoned him did it covertly, like a spy—not like a soldier.

Though poison and treachery have common roots in the history of warfare, the connection seems to have been downplayed by modern commentators and manuals of legal advice for the military. One exception appears to be Air Force pamphlet 110–31, *International Law: The Conduct of Armed Conflict and Air Operations*, which refers to chivalry as one of three pillars of the international law of armed conflict and defines chivalry as a prohibition against treacherous misconduct and the use of poison.[28]

The emphasis in the contemporary military manuals seems to be on the danger of poisoning the food and water supply. According to the British military manuals, "Water in wells, pumps, pipes, reservoirs, lakes, rivers and the like, from which the enemy may draw drinking water, must not be poisoned or contaminated."[29] Lieber was as deeply committed to this view as he was to the association between poison and perfidy. As his code posits in Article 70, "The use of poison in any manner, be it to poison wells, or food, or arms, is wholly excluded from modern warfare. He that uses it puts himself out of the pale of the law and usages of war." Significantly, Lieber's explicit ban on poison and perfidy in Article 16 was repeated word for word (without attribution) in the U.S. military manuals up to the end of World War II. In the 1956 army manual entitled *The Law of Land Warfare*, the language shifts to the following commentary on the prohibition against poison in Article 23(a) in the Hague Convention:

> The foregoing rule [banning poison] does not prohibit measures being taken to dry up springs, to divert rivers and aqueducts from their courses, or to destroy, through chemical or bacterial agents harmless to man, crops intended for consumption by the armed forces (if that fact can be determined).[30]

Implicitly, the ban on poison prohibits measures that affect the water and food supply in a way that is harmful to human beings. This adds a different twist. The argument is not only that using poison against enemy troops is perfidious and dishonorable, but that using poisonous means creates great risks for the civilian population. Moral philosophers could easily accommodate this shift in emphasis, but they have trouble engaging with the black letter rules of the law of war, preferring instead to deal with the moral principles that are familiar from their own philosophical tradition.

Philosophers think about abstract moral principles without paying attention to the problems of applying the norms in practice. They rarely discuss how to enforce these norms, either because they assume that human beings will always follow them or because they regard questions of application as being needlessly empirical. Indeed, this attitude has a long tradition in philosophy. Kant could favor the death penalty without thinking for a minute about the institutional problems of applying it without error or undue cost or delay. He would be shocked to witness the practice of capital execution in the United States—a process that requires numerous appeals and forces the condemned to wait on death row for years, even decades, for a resolution.

The notion of institutions includes certain architectonic distinctions that structure the law of war. The most basic of these is the radical separation of jus ad bellum and jus in bello. The lawfulness of the war has no bearing on the proper conduct of warfare. As a soldier, regardless of whether you are part of the aggressing or the defending army, you may not violate the strict prohibitions of the Geneva Conventions or the Rome Statute. Though the Latin phrases are arguably of recent vintage, Kant made exactly the same distinction in German in 1797.[31] But however well entrenched this distinction is in the law of war, philosophers still balk at it. A good example is Nagel, who argued in 1972 with the Vietnam War in mind, "If the participation of the United States in the Indo-Chinese war is entirely wrong to begin with, then that engagement is incapable of providing a justification for *any* measures taken in its pursuit."[32] In other words, if the war is unjustified and unlawful, everything done in the course of that war is unjustified and unlawful. This is a shocking position. If it were true, and if one believed that the war in Iraq is wholly unjustified, then every act of killing the enemy in Iraq would constitute unjustified homicide.[33]

This way of thinking—disregarding the distinction between jus ad bellum and jus in bello—accounts for the curious use of language in the work of Jeff McMahan. Following Nagel, McMahan refers to the soldiers who fight in an unjust war as "unjust combatants."[34] Thus every soldier doing his or her duty and refraining from committing prosecutable war crimes still warrants the label "unjust." This reflects a bias against the military, which we explore in the next section. McMahan's position draws on the work of the late great liberal philosopher John Rawls. In his classic

1971 work *A Theory of Justice*, Rawls initially took the position that the justice of the war should have a direct bearing on whether a soldier has a right to kill.[35] But later in the same decade, Michael Walzer took a step back from this disconnect between the two dimensions of just war and argued in favor of a presumptive distinction between jus ad bellum and jus in bello, a once well-known distinction from just war theory that had long since fallen out of usage. The only exception, according to Walzer's influential theory, should be cases of "extreme emergency." He had in mind the necessity of defeating National Socialism in World War II. Rawls came around to this position in his later work, but some contemporary philosophers, such as McMahan, steadfastly hold on to the notion that all killing in an unjust war is prohibited.

Why should philosophers be reluctant to accept such a conventional truth about the law of war? As a matter of abstract moral principle, Rawls might have been right in 1971. But lawyers think differently because they are concerned about application of the principle over the long run, in a series of factually disputed cases. If the influence of jus ad bellum on jus in bello were an open question, no soldier would ever know for sure what the rules of warfare are. The scope of lawful behavior would depend on a determination of which nation was the aggressor and which was the lawful defender. This uncertainty would be unfair to individual soldiers, who risk criminal liability if they make a false judgment about the lawfulness of the orders they execute.[36]

With only a few exceptions, philosophers rarely deal with how moral norms will be applied and enforced in real life. Or, to the extent that they recognize the problem, they regard it as a matter to be resolved by another academic discipline, such as applied ethics, not philosophy proper. But lawyers cannot even begin to think about norms of conduct without contemplating their application in practice; if an organizational principle in the law of war—such as a flexible and moving boundary between jus ad bellum and jus in bello—does not lend itself to easy application, this is of immediate concern to lawyers. For lawyers, norms enjoy a suspended existence in the abstract, but they come to life in institutions, under the pressure of adapting the norm to real-life situations. Institutions still seek justice in every case, but those within the system understand that perfection is not always possible. Mistakes will be made. The criminal trial is a good example. It requires proof of guilt beyond a reasonable doubt, not proof beyond "every conceivable doubt."[37]

The reason for adopting a rigorous distinction between jus ad bellum and jus in bello is the need for a bright-line cleavage that is workable in the field of battle. Soldiers do not have to think about who started the war. They know that, regardless of who started the conflict, certain means of warfare are clearly illegal. Regardless of who the aggressor is, no one may target innocent people, employ poison, use dumdum bullets, or declare that no quarter will be given to prisoners.[38] These and a host of other provisions govern battles in the field. The world within the war—insulated by an indifference to who started the war and why—has its own law. It is a law for both sides, and those who refuse to follow this law are liable for war crimes under the Rome Statute.

The institutional perspective of lawyers leads to an entire apparatus of rules that are alien to philosophers. These are the common law rules of evidence, the presumption of innocence, and the privilege against giving testimony and the constitutional rules against using certain forms of evidence. One could say that lawyers are, by their nature, pragmatic about norms. They want to know not only whether a norm is right, but whether it will work in practice.

Kant is one of the very few philosophers to appreciate this style of legal reasoning. In a dramatic and underappreciated passage, Kant explains the difference between abstract norms and how they work in practice. He gives four examples where the norms in the real world—what he calls, in translation, "Right in the civil condition"[39]—differ from what he calls the abstract theory of Right. His first example is the promise to make a gift. In theory, the promise should not be binding, because there is no reason to assume that the promisor meant to surrender his freedom gratuitously. Yet in practice, courts can assume that the promisor meant to be bound "because otherwise its verdict on rights would be made infinitely more difficult or even impossible."[40] For the sake of practical administration, then, courts can adopt a rule that is just the opposite of what should be true in principle.

Kant makes the same point about the hypothetical case of being caught in the rain and borrowing a coat from a stranger.[41] If the coat is stolen from the borrower (who has not paid for the use of the coat), who should bear the loss? In a state of nature, Kant reasons, the borrower should bear the risk, but "before a court," according to "Right in the civil condition," the lender should know that the burden is on him to stipulate the conditions on which he expects to receive the coat back. Kant's reasoning is a bit clearer in the third example, a sincere purchaser who buys from someone who deceitfully represents himself as the rightful owner of an object.[42] In the abstract theory of property, no one can acquire title from someone who does not have it, but "in the civil condition," a court would decide differently. If someone buys a horse in good faith on the open market, then he should acquire not only a personal right against the seller but also the property in the horse—even if the seller was not the actual owner.[43] In the final example, Kant defends the use of oaths to induce people to be honest in their testimony. He regards oaths as a form of "superstition," but he endorses their use on the grounds that they are "expedient."[44]

These concessions to the necessity of practical justice illustrate an important point. Kant had a vivid understanding of the difference between law and morality. The administration of the law in the world requires compromises with abstract principle. It would be better in theory to protect the freedom of promisors, lenders, owners, and witnesses against the burdens of conflict in the courts. But in practice, we hold people to their promises;[45] they must take the risk of lending objects without securing their rights; we tolerate owners losing their rights in good-faith commercial transactions; we require witnesses to take oaths. In the same spirit, we should enforce a strict separation between jus ad bellum and jus in bello. In most cases, determining who is aggressing and who is defending is open to debate. As Kant reasoned, judges in court or soldiers in the field should be able to make judgments on the basis of what is certain. The rules in the law of war are difficult enough to apply, and the burden of false judgment carries serious consequences for soldiers who obey an unlawful order—or who disobey lawful orders. From either a moral or a legal point of view, there is no warrant for making these decisions more difficult by requiring soldiers in the field to assess the lawfulness of the war in which they are engaged.

B.   *The Liberal Bias*

Philosophers in the English-speaking world are by and large liberals. Following the work of Rawls, Ronald Dworkin, Bruce Ackerman, and Joseph Raz, they think primarily about individuals as the units of moral action. Thus they have a hard time understanding the basic institutions of the law of war, which are essentially collective in nature. At least, this is the way we thought about the law of war in its formative stages. Lieber captured the spirit well by conceding that "the citizen or native of a hostile country is thus an enemy, as one of the constituents of the hostile state or nation, and as such is subjected to the hardships of the war."[46] Since the adoption of the Geneva Conventions in 1949 we have recognized a more rigorous distinction between combatants and civilians, but at least we accept that all combatants are part of the enemy at war and all should be engaged in the arena of hostilities.

The implication of an arena of warfare is that all participants within it are permitted to kill and all are subject to being killed. It makes no difference whether they are on the front line or providing backup logistical support. It has no legal bearing whether they are aiming to kill another or whether they are asleep. They are all subject to the rules of jus in bello.

Philosophers disregard this foundational principle in the law of war to a remarkable extent. Nagel set the tone for the modern philosophical literature by wondering whether it is permissible to kill enemy soldiers when disabling them might be sufficient for self-defense.[47] His starting point is that the justification for the use of deadly force must be the harmful threat emanating from the enemy. McMahan continued this pattern by labeling as "materially innocent" all those who do not pose "an imminent threat of harm."[48] David Rodin is the latest in this line of philosophers who collapse collective and individual self-defense.[49]

Indeed, one can see further evidence of the schism on this point between lawyers and philosophers by carefully examining their use of the terminology. For most philosophers, "collective" self-defense refers to the self-defense exercised by the nation as a collective, through its army, against its enemies. But international lawyers mean something completely different when they talk about collective self-defense. To them the term refers to a group of nations that fight to protect their common interests, much as the Allies did in World War II or as one finds in regional defense pacts such as NATO.[50] What philosophers call collective self-defense (i.e., the defense of the nation) is simply called "individual self-defense" by the international lawyers, a term that philosophers reserve for self-defense exercised by individual human beings. Philosophers are inclined to think of national self-defense as collective because they analyze the army with the same categories that they use to describe criminal gangs. The only way they can think about enemy soldiers is the way they might think about a band of robbers attacking civilians.[51]

Once they take this fundamentally wrong theoretical turn, philosophers have a hard time explaining why soldiers are not guilty of homicide when they kill other soldiers who are not actually engaged in attacking them. They must speculate about the implicit coercion of military service or other implausible theories to explain the well-accepted distinction between routine military operations and war crimes. Their moral contortions on this issue would come as a great surprise to lawyers, if the latter group ever bothered to read moral philosophy. For lawyers, the immunity of combatants for killing combatants is taken for granted. They understand intuitively that war is a collective engagement and that when combatants don a uniform and become subject to a chain of command they are no longer protected persons under the Geneva Conventions.

If moral philosophers are confused about whether a combatant may kill other combatants with immunity, they are also in a quandary about civilians who kill soldiers. Because they do not grasp the immunity of combatants, they do not understand that *only* combatants enjoy immunity conferred by the law of war. Even as the 1977 Protocol I to the Geneva Conventions extends the principles of armed conflict to armed resistance "against colonial domination and alien occupation,"[52] the principles of the Geneva Conventions still apply. These principles presuppose that those engaged in warfare are not protected persons; that is, they are not the sick, the wounded, prisoners of war, or civilians.[53] Interestingly, there is no specific crime in the Rome Statute that covers incidents of civilians attacking soldiers.[54] This absence is telling, since the Rome Statute includes specific provisions for all other categories, including one provision that holds commanders criminally liable for failing to supervise their soldiers.[55] Also, the Rome Statute eliminates immunity for heads of state, who are typically civilians rather than military personnel.[56] But for routine assaults committed by *civilians* against soldiers, the more appropriate response is prosecution under the applicable domestic law, not the Rome Statute.

The total disconnect between the philosophical and legal literature means that neither side has felt it necessary to justify its assumptions. So far as we can tell, the philosophers have not mounted a serious argument against the collective conception of military forces, and the lawyers never bothered to refute the moral quandaries about why soldiers enjoy immunity for actions committed in the ordinary course of fighting. The philosophical view derives from the principles of liberal individualism that are dominant in the academy today. The more intriguing question is what the lawyers could argue on behalf of their assumptions about jus in bello.

The law of war has developed by justifying exceptions to the general principle that nations engage in war with the entirety of other national entities. Lieber began this process and the Geneva Conventions completed it by refining the exception for "protected persons," who are protected from the theater of hostilities by virtue of their status as noncombatants. The moral philosophers now turn the question around and demand a justification, not for the exception, but for the basic rule that combatants enjoy immunity for killing combatants. Generating this justification is more difficult than meets the eye.

As argued elsewhere, we certainly believe that the Romantic sensibilities of the late eighteenth century and early nineteenth—the sweeping ideas of Rousseau, Herder, Hegel, and Fichte that emphasized the collective "self" of a people that found expression in a particular state—could justify the traditional view that nations go to war with nations.[57] However, no one is obligated to adopt Romantic rather than liberal (and individualistic) premises. Believing in the collective action and the collective guilt of the nation may be too much of a stretch for many serious thinkers nurtured by individualistic ways of thinking that grew out of the Enlightenment. Yet even more difficult is reconciling the Romantic vision with the conception of protected persons developed by the Geneva Conventions.

Romantics think of nations as actors in history. The nation has a culture, a history, usually a common language, often a unifying religion. The army is but a subset of the nation. Lieber could grasp the idea that nations go to war, and he therefore recognized that civilians must share in the "hardships of the war."[58] Yet he pleaded for a general principle of minimizing harm to "unarmed citizens."[59] We are not sure that the foundations of the law of war have advanced much beyond this point.

To account for present assumptions in the law of war, we need a two-stage argument. First, we must explain why the entire nation goes to war; second, we must justify the exceptions for protected persons under the Geneva Conventions. We tackle both of these conceptual problems in this book.

5. HISTORY IN A NUTSHELL

Every legal system has a concept of self-defense; the contours of the idea have engaged the imaginations of lawyers and philosophers for thousands of years. The Bible acknowledges the principle of self-defense when it recognizes the right of the homeowner to kill the "thief breaking in."[60] If the homeowner kills to protect himself, he has no blood on his hands.[61] He carries no "blood guilt," as Kant called it,[62] and therefore there is no need for punishment or any other social response. The Talmud supplemented this core idea by adding a right to use defensive force against aggressors threatening human interests. The rabbis reasoned that it was permissible to kill a *rodef*, or aggressor, who was threatening to kill, rape a man, or commit adultery with a betrothed woman. It was not enough for the immediate use of force that the aggressor threaten to commit a nonviolent capital offense, such as desecrating the Sabbath or engaging in idolatrous practices.[63] These nonviolent crimes constituted great evils, but they did not threaten irreparable harm to a human being, and they were not time-sensitive, which meant the community could afford to wait for a trial. Thus there was no right of immediate resistance. These nonviolent crimes were still subject to the death penalty, but only at the hand of a court or divine judgment.

There are many problems in the Talmudic account of self-defense, as we shall see, but one that stood out in the early history of moral philosophy was the kind of intention required for self-defense. Thomas Aquinas was particularly concerned about whether self-defense actually authorized the killing of human beings, and to avoid arriving at the conclusion that it did, he invented a powerful moral doctrine called "double effect."[64] The purpose of killing in self-defense, he reasoned, was not to kill per se but to ward off the attack. The use of force had to be directed against the attack, not the attacker. The death was a side effect of the legitimate purpose rather than the goal itself.

Many moral philosophers, drawn to the theory of self-defense in the past few decades, speak freely about a right to kill in self-defense. The Oxford philosopher John Gardner argues that the moral reasons for avoiding killing are suspended in cases of self-defense.[65] His reasoning is ingenious, but it fails to make the important Thomistic distinction between the right to kill and the right to ward off the attack. As we shall see, several characteristics of self-defense—most notably, the requirement that the force be necessary—will lend themselves to explanation only if we recognize this vital distinction.

In his sophisticated account of self-defense in both international and domestic law, Rodin argued that self-defense should be considered a "liberty," used as a term of art in a framework designed by Wesley Hohfeld to account for interpersonal rights and duties.[66] To call self-defense a liberty in the Hohfeldian sense is to say that neither the original attacker nor anybody else has a right to prevent the defender from using force. Unfortunately, this tells us very little about the nature of the "right" to use force; we have an infinite number of liberties, ranging from going for a walk to speaking our mind about public issues to engaging in religious worship as we please. Hohfeld's theory would offer us very little guidance if it merely led us to think that self-defense was just another of our many constitutional liberties enjoyed in modern democratic societies.

The Hohfeldian scheme is actually designed to describe interpersonal relationships in private law. One agent bears a right, and so another is endowed with a correlative duty. If A has a right against B, then B is under a duty to A. The Hohfeldian scheme can also account for legal relationships. If A has a power to exercise a contractual option against B, then B is under a liability relative to that power. The idea of a liberty is simply the negation of the right/duty relationship: no one has a right against B with respect to a certain action, and therefore B has a liberty to engage in that action.

Rodin presupposes, as do many moral philosophers, that the right of self-defense arises in an interpersonal relationship between two specific individuals: the aggressor and the victim or defender. The scope of self-defense depends on the way the two interact—on the parties' relative innocence or fault or on the way one has coerced the other. If there is a right to self-defense, therefore, it should be understood as A's interpersonal right against B. Because the right applies to B, it is appropriate in the idiom of this philosophical school to speak of a "right to kill" in self-defense. This idea is tempting and obviously seductive, but fundamentally flawed.

Rather, to vindicate Aquinas's concern about the taking of human life, the proper characterization of self-defense is based not on a right to kill, but on a right to ward off the attack. If the aggressor must suffer as a result, it is a regrettable but unavoidable side effect. This is the way lawyers think about self-defense, both in the domestic sphere and in international law.[67] To understand the lawyers' point of view, we must review Kant's theory of self-defense in his *Legal Theory* of 1797. Kant begins on the assumption that under the law, understood as an ideal set of principles, everyone is entitled to a maximum sphere of external freedom. This ideal legal system is defined as the Right, or the Law as a set of principles of "ordered liberty."[68]

Aggressors violate this guaranteed sphere of individual freedom. Because the violation of the Right (Law) is also a violation of personal rights, in every case the aggressor threatens not only an individual right, but the legal order as a whole. The Right (Law) encompasses all personal rights—to life, to property, to privacy, to human dignity. This bundle of rights is often referred to as autonomy. Under a proper legal order, all aspects of personal autonomy, or external freedom, are protected against aggression.

When autonomy is violated, the victim requires a remedy. Suing in a court of law is an appropriate remedy in many cases, but not in all. The first available remedy is to use force to prevent the aggression, thereby preventing the violation of the Right; but this remedy should be available only when the wrongful threat is still pending and it can still be avoided. The key word here is "necessity." The defendant is allowed to do all that is necessary to prevent the violation of his personal autonomy. If the violation has already transpired, however, then the only lawful response available is to bring a lawsuit to repair the wrong.

Whether the former should be available in all cases of aggression raises a different issue, with a different label, namely, "proportionality." There might be some cases in which the victim/defender must hold back and not take remedial action by using force. When and if this restraint is required poses difficult questions, which we take up in due course in chapter 4.

This is Kant's argument for a right of self-defense: a remedy designed to prevent violations of the external freedom that is protected under the principle of Law as guaranteed freedom and ordered liberty. The key to thinking about the right of self-defense, then, is not the relationship with the aggressor but the autonomy of the victim/defender. The defender is morally permitted to do what is necessary to restore his autonomy (or external freedom). To do so, however, he may not necessarily have to *kill*. He need only use enough force to repel the attack. The idea that self-defense unequivocally generates a "right to kill"—something you might even enjoy doing—is a perversion of a great legal tradition.

The Kantian method of focusing on individual autonomy replicates, in domestic law, our vision of the independence of nations. It is not far-fetched to picture nation-states as islands, each with an absolute right to control what occurs within its territory. When a nation-state suffers aggression, it may repel that violation of its rights by force. Similarly, if we think of every individual as an island, then, by analogy to international law, domestic law entails a right to repel aggressors who trod upon the space of the individual.

The Kantian approach gives due regard to the aggressor's right to life by instead focusing on the victim's right to restore his autonomy. There is no right to kill—only a right to take the measures necessary to prevent violations of the defender's autonomy. This Kantian approach to self-defense has generated a pattern of legal discourse on the Continent that is distinctly different from the prevailing approach in English-speaking countries. German codes and literature have always referred to the right of self-defense as *Notwehr,* or necessary defense, implying that more interests and rights are protected than just the "self."[69] In the Kantian conception of the Right, all aspects of personal autonomy are protected by this right of "necessary defense." This notion of necessary defense will play a key role in our explorations in subsequent chapters as we seek a new interpretation of self-defense in international law.

2

HOW TO TALK ABOUT SELF-DEFENSE

1. THE GRAMMAR OF SELF-DEFENSE

Diversity reigns in the domain of self-defense; we have variations suitable for all theoretical tastes. Our only disagreement derives from the genuine difficulty of understanding what goes on when we justify killing in response to aggressive force. As we work through these variations in the realm of domestic law, we shall discover new byways of thought to explore in international law.

A. *Self-Defense as an Exception*

Criminal law theorists are often tempted to think of claims of justification as exceptions to the norm that establishes a basic wrong or prohibition.[1] The reasoning goes something like this: Norms establish the wrongs—homicide, theft, and so on—that apply in standard, everyday life; however, in exceptional cases you might be justified in committing one of these wrongs on grounds of self-defense or necessity. The proper way to think of claims of justification, therefore, is that they are exceptions that should be stated as negative elements of the norm. As the legal scholar Michael Moore suggests, think of Moses descending the mountain carrying two tablets. The left-hand tablet contains the prohibitions of the criminal law; the right-hand inscriptions indicate the exceptions that permit the violation of the left-hand norms under certain circumstances. To understand the norm as it actually functions in society, you have to look at both hands at the same time.

A good example is larceny, defined in the common law as containing the following elements in the left-hand tablet:

1. An act of taking
2. an object
3. that belongs to another
4. taken from the possession of another (not necessarily the owner)
5. with the intent to deprive the owner permanently of possession.

Recently, legislative drafters have started listing the elements of defined offenses in this manner. You find it in the Elements of Offenses defined by the Assembly of State Parties to clarify the definitions of genocide, crimes against humanity, and war crimes in the Rome Statute; you also find it in the sophisticated document called *Military Commission Instruction Number 2*, published by the U.S. Department of Defense.[2] Note that these elements say nothing about possible claims of justification or excuse; they refer only to the paradigmatic or standard cases of wrongful larceny. Claims of justification and excuse are stated in the right-hand tablet, and they arise, as we have noted, only in exceptional cases. Though self-defense would be rare in larceny cases, imagine the following scenario. One street gang is attacking another by using shoulder-fired missiles, and the least costly way for the defending gang to protect itself is to seize the weapons from their enemies. They need not kill them or even detain them. Once deprived of their weapons, the aggressive gang can do little to harm their adversaries. This would be a case of self-defense justifying an act of larceny, namely, depriving the aggressive gang of their missiles permanently.

Self-defense, then, would be an exception, stated on the right-hand tablet. Moore's point is that the overall norms that define larceny require paying attention to both tablets. You cannot read the left tablet without the right. The complete set of elements would read:

1. An act of taking
2. an object
3. that belongs to another
4. taken from the possession of another (not necessarily the owner)
5. with the intent to deprive the owner permanently of possession.
6. The elements of self-defense are not satisfied.[3] (In short, "No SD.")

But this is not the only possible exception to the norm defining larceny. Here are a few more:

7. Necessity (i.e., stealing to save the life of a starving person)
8. Kleptomania or other forms of insanity
9. An order of eminent domain by the state to seize the property

It is hard to know when we have exhausted the potential list of justificatory claims, so courts have the capacity to recognize new claims of justification in the course of litigation. This is the lesson of a famous German case decided in 1927, in which the Supreme Court for the Weimar government recognized a new justification based on the balancing of the competing interests of fetus and mother in cases of abortion.[4] Today, the Rome Statute explicitly recognizes the possibility of supplementing the defenses recognized by statute. Article 31(3) specifically authorizes the court to consider additional grounds for excluding criminal liability by looking to additional sources of law. Article 21(1)(c), which governs the court's use of national law, states that the court can look to "general principles of law derived by the Court from national laws of legal systems of the world."[5]

Also, Article 21(1)(b) authorizes the court to consider general principles of international law when making its decisions. This latter point raises a tricky issue of interpretation. Some scholars, including the distinguished Antonio Cassese, are inclined to view this provision as allowing the court to use "customary international law," or general principles of international law based on state practice that are not necessarily codified in any treaty or statute.[6] There are dangers lurking in this analysis. The Rome Statute does not make an explicit distinction between the definition of a criminal offense (which imposes criminal liability) and the recognition of a defense (which excludes criminal liability). While it is fine to invoke customary law when recognizing a defense, it is quite another thing to invoke it when interpreting the definition of the offense itself. The bedrock principle of criminal law, *nulla poena sine lege* (no punishment without prior legislation), stands for the idea that the portions of the law that generate criminal liability for individuals must be written down in advance.

In other words, the International Criminal Court may recognize new claims of justification, even though this principle of legality—nulla poena sine lege—would prohibit it from developing new common law offenses.[7] This is a fundamental difference between prohibitions and claims of justification. The elements of the prohibited act are "closed," as required by the principle of prior legislative definition of offenses. The list of justifications, however, is not closed. The court is free to add new ones as justice requires.

This crucial difference between the affirmative elements of the prohibition and the so-called negative elements led the philosopher H. L. A. Hart to suggest, at one stage of his writing, that one could state the basic norms of criminal law as necessary and sufficient conditions for liability.[8] Although the necessary conditions are laid out in the statute defining the crime (e.g., killing civilians as a war crime), there is no closed list of sufficient conditions for *defeating* liability. The left tablet contains a complete list of prohibitions, but the exceptions offered by the right tablet are incomplete. It is always possible that a new claim of justification, previously unforeseen on the horizon, could destabilize the appearance of a well-stated set of affirmative and negative conditions of liability.

An alternative to viewing self-defense as a negative exception is to consider claims of justification as a separate set of norms that, as they were once called in common law pleading, "confess and avoid" the prohibition of the offense. At common law, a defendant could plead "confession and avoidance" by admitting the basic facts alleged but present new evidence to avoid liability, as would be the case if a defendant admitted to a killing but argued self-defense. He would, in a sense, be confessing to the basic facts of the killing but avoiding liability for some additional reason. To confess and avoid is to perform, in a sense, an end run around the prosecution's argument: the defendant faces the facts squarely and admits to them but argues a path that the prosecution has not covered in its allegations. In our example of taking the missiles from the street gang, these elements are obviously satisfied. The street gang threatens an imminent attack with their missiles; the response of taking them is necessary and proportional, and the defenders act with the intention of defending themselves.

The language of confession and avoidance has largely disappeared from the criminal law, relegated to the dustbin as archaic terminology from common law. This is unfortunate, because the language is useful for a proper understanding of legal justifications. To extend the metaphor of the two tablets, a claim of justification is not simply the other side of the norm, the right hand offsetting the left hand. Rather, justificatory claims of self-defense, necessity, and the like arise on a different plane of analysis. They do not negate the norms of prohibition but provide arguments of confession and avoidance.

Many important legal consequences follow from this view that claims of justification are grounded in a separate norm and fully independent of the rule that defines the offense. The most important of these implications is that claims of justification are still compatible with a perception of the rule as having been breached. That is, in our hypothetical street gang case, the defenders do indeed commit larceny—they have infringed upon the norm prohibiting permanent taking from another—but they do not act wrongfully, because they are justified in taking and keeping the missiles.[9] It is a critical feature of the relationship between prohibitory norms and the countervailing norms of justification. The infringement or transgression against the rule that defines the offense remains intact even if, over all, the conduct is justified.

John Gardner searches for the proper way of expressing the nature of justification in light of his commitment to the plausible view that legal prohibitions provide reasons against engaging in the conduct.[10] Norms of justification counteract these reasons, but not by providing stronger reasons for engaging in conduct. If they did, those who kill in self-defense would not only be permitted to act, they would be *obligated* to follow the stronger reasons in favor of killing. Instead, the reasons for not killing remain intact, but the defender is nonetheless permitted to act. We are not sure an adequate account of this legal permissibility is available if we adhere to the idea that the act prohibited (killing) is the same as the act permitted (killing). The better way, suggested by Aquinas's theory of double effect, is to recognize that a proper analysis requires two different *descriptions* of the act in question. When an act is justified, we use a different description of the act to talk about its justification than we do when we define the prohibited act in the first instance. The prohibited act is killing a human being; the justified act is thwarting the attack. One is permitted to stop the attack even though the indirect consequences to the aggressor might be fatal.

This view is critical to our understanding of self-defense. And if we hold to it, we cannot accept Moore's metaphor of the two tablets of norms, some positive and others negative. We must recognize the wrong of killing at the same time that legitimate defense is treated as permissible. This might seem like a purely metaphysical issue, but in fact it is of critical importance in shaping our perceptions of aggression and justification, both in the domestic and in the international spheres. Here is another way of putting the point: Justifications such as self-defense negate the invasion of the legal interest (such as life) protected by the rule or norm defining the offense (such as homicide). But the erasure is not total, for it leaves a shadow of this invasion. The justification, as it were, shines a light of rectitude on the harm done, thus casting the conduct in relief as lawful and free from stigma, even as the shadow of the invasion remains.

The shadow of the violation serves to define the relevant legal interest. In cases of larceny, the legal interest is the possession of property; in rape, it is sexual autonomy; in homicide, the relevant legal interest is life itself. These interests are violated even if the conduct is justified. If, for example, Dan the defender kills Alice the aggressor in self-defense, Alice's right to life is infringed, but the infringement is, all things considered, justified. While the shadow of the invasion is still perceptible—in this case, the killing of a human being—the principles of justification are superimposed on the invasion. If the claim of justification were integrated into the norm as a negative element, then it would function in the same way as a denial that any violation had occurred at all.

Let us see how that would work for a particular offense such as larceny. The definition of the offense would look like this:

1. An act of taking
2. an object
3. belonging to another
4. from the possession of another
5. with the intent to deprive the owner permanently of possession.
6. No SD (without the justification of self-defense).[11]

If the justification completely negates the invasion caused by the offense, then all elements from this list are erased. So there is no act of taking, for example; there is no violation of the norm, and not even the shadow of a violation. The same would be true, according to this model, in all cases of self-defense, regardless of the offense in question. If the elements of self-defense are satisfied, there is no violation—no shadow of a wrong. Thus, if self-defense is merely an exception, a negative element of the definition, then no rule is violated when the defender acts to save his own life.

If we were to accept this model, we would have to figure out what interest is being protected by this legal norm. This is not so easy to do, and the only possible legal interest would appear to be the rights of non-offending persons. Specifically, if self-defense in larceny or homicide were a negative element of the rule prohibiting the crime, then the interest in question would not be the right to possession of property or life in general, but rather the right to life or property possession under *certain conditions*, namely, when the defender does not have a valid claim of self-defense.

In effect, only non-aggressors would be protected. Aggressors would have no rights at stake; their legal interests would not be violated or infringed upon by legitimate and justified actions in self-defense. This is conceptually unsound.

The view that only non-aggressors are protected against the use of force leads to the implausible conclusion that aggression itself effects a forfeiture of the right to life—or, in the case of states, to the loss of their political independence. This would mean that aggressors become, in effect, outlaws, totally unprotected against violence. If this were true it would be extremely difficult to explain the six elements of self-defense that we outline in chapter 4. In particular, the required elements of self-defense—an overt attack, imminence, necessity, proportionality, and intentionality—would lose their grip if no protected interest remains in cases of self-defense. These elements of self-defense do not stem from the outlaw status of the aggressor, but from the particular nature of acts of aggression and how they might be legitimately be opposed.

These reflections have a direct bearing on the structure of international law. The norms prohibiting larceny or homicide are analogous to the prohibition against war in the United Nations Charter Article 2(4):

> All Members shall refrain in their international relations from the threat or use of force against the territorial integrity or political independence of any state, or in any other manner inconsistent with the Purposes of the United Nations.

The only exception to this—apart from military action authorized by the Security Council—is Article 51, which recognizes the inherent right of self-defense (or le droit natural de légitime défense). What, then, is the relationship between the prohibition in Article 2(4) and the inherent right recognized in Article 51? In considering this question, we can profitably draw on debates in the domestic theory of self-defense.

Is there a protected legal interest underlying Article 2(4)? We would like to think that it is either the "territorial integrity" or "political independence" of the states constituting the United Nations. This neat and coherent interest of states would be analogous to the individual's right to life and property. The values of integrity and independence are as compelling when applied to states as they are in expressing the dignity of individuals. In another age, one would have referred to this autonomy as the "sovereignty" of states, but the UN Charter expresses the spirit of modernity by avoiding the concept of sovereignty, except in two references to the "sovereign equality" of all states. Though political theorists still talk at great length about sovereignty, the concept has lost its utility to lawyers and constitutional drafters.[12]

The idea that all states are of equal value underlies the prohibition of both force and the threat of force in Article 2(4). No state can assume the authority to dictate policy to another. If some states—those, say, with liberal democracies—were morally superior to others, they would naturally claim a right to interfere in the politics of oppressive regimes; they would regard themselves as authorized to decide when there should be a "regime change" for the sake of human rights or democratic politics. Once the principle of moral superiority is admitted, however, oppressive fundamentalist regimes could plausibly invoke the same principle of intervention to advance their "true religion."

Would that the references in the Charter to "equal sovereignty" could support as strong an analogy between the autonomy of individuals and the autonomy of states, but one clause in Article 2(4) stands in the way. The protection of states and their autonomy is modified by the catch-all phrase "inconsistent with the Purposes of the United Nations." Force that is consistent with these vaguely defined purposes would presumably be exempted from the prohibition imposed by Article 2(4). Some writers have invoked this clause to defend various forms of urgent and temporary intervention in the internal affairs of other states. For example, Israel used this argument to justify its intervention in Uganda to rescue its nationals being held by hijackers at the Entebbe airport in 1976.[13] If the Security Council could not intervene, then arguably Israel had the right to rescue its nationals on its own.

The 1949 condemnation by the International Court of Justice (ICJ) of the United Kingdom's intervention in Albania's Corfu Channel to recover evidence tends to support a stricter view of territorial inviolability of all states.[14] Yet domestic legal systems have had to struggle with the same problem; in cases of emergency, such as the Entebbe incident, it is unacceptable for the owners of the land to refuse access to their property when it is necessary to save life and property.[15] For example, if it is necessary to moor a ship to avoid the dangers of a storm, the littoral landowners (the owners of the dock in question) must yield and accommodate those in distress. If the sanctity of land gives way to the sanctity of life in these cases, why should Uganda not be required to tolerate and accept Israeli intervention to free the hostages of a hijacking? Exploring the analogy between international law and domestic law raises such unanswered questions.

The typical case for humanitarian intervention is also based on a reading of Article 2(4) that acknowledges the necessity of temporary violations of territorial sovereignty to prevent the loss of life. A controversial example is NATO's 1999 bombing of Serbia in an effort to prevent ethnic cleansing in Kosovo. Apparently, Belgium has made the most articulate case for intervention, based, intriguingly, on a duty to avoid the mass killings of innocents.[16] If all states have a duty to avoid mass casualties, then they also have a right to intervene to perform that duty. Later in this chapter we consider the generalized case of duty as the foundation of the right to use defensive force. Admittedly, the bombing of population centers in Yugoslavia was a rather brutal form of humanitarian intervention, which went far beyond a temporary incursion comparable to the Israeli commando-run intervention in Entebbe. Yugoslavia complained before the International Court of Justice that the NATO bombing was a disproportionate response to Yugoslavia's actions against the Albanians in Kosovo.[17]

Unfortunately, it is difficult to define the basic legal interest being protected by the international legal prohibition against aggression. There might even be a good argument for the view that aggressors forfeit their interest and their territorial inviolability when they engage in armed attacks. This would be the result of incorporating Article 51 into Article 2(4) as an exception to a single norm, which would read, in essence: "The territorial integrity and political independence of all member states shall be secure against intervention by force unless that intervention is exercised in self-defense (legitimate defense) by a member state to protect its own territorial integrity and political independence."

Occupation also raises puzzling issues. Take Israel's occupation of Egyptian, Jordanian, and Syrian lands after the 1967 War, or the Allied occupation of Germany and Japan after World War II. What should an occupying country do after successfully repelling an aggressor? Should it simply withdraw its forces? If we follow the teachings of domestic self-defense, we would have to say yes. If the defender Dan disables the aggressor Alice, takes her weapons, and sits on her long enough to make sure that the attack is neutralized, he still must eventually get up and let Alice go. He cannot hold her prisoner permanently to ensure that she never attacks him again. If the same principle held in international armed conflict, the United States and Israel would have to withdraw as soon as they disarmed their enemy; they would not be allowed to occupy conquered lands until they were satisfied that the defeated nation was willing to sign a peace agreement or make a proper commitment to mutual recognition and to living as neighbors without threats of invasion.

There is, of course, a radical difference between the situation just described and an occupation resulting from an aggressive war, as in the case of Iraq's occupation of Kuwait in 1990. In such cases, the international community must insist—even to the point of using force—that the aggressor terminate the occupation immediately. In the case of a defensive war, the conventional international response is to tolerate and even support occupation until a peace treaty is negotiated and signed. Even the Fourth Geneva Convention implicitly recognizes the legality of occupation by prescribing the conditions for its exercise.[18]

How do we explain this dichotomy in international practice between aggressive and defensive occupation? Aggressive occupation—by Germany in, say, Poland—is arguably a continuation of the initial invasion and conquest. The defensive occupation is an extension of the right of self-defense. This, as we noted earlier, is the modern view. The traditional approach to occupation stresses the rigorous distinction between jus ad bellum and jus in bello and suggests that the postwar question of occupation should be insulated against the criteria that determined the justification of the war in the first place. Occupation is, arguably, the same regardless of how the war began. Just as the status of being the aggressor in an international conflict does not make killing enemy soldiers a war crime, so, too, it should not change the legal standards for how an occupation should be evaluated after the cessation of hostilities. If jus in bello depended on who was the aggressor, nations could never agree on the rules of warfare because they would be engaged in a constant argument about who really attacked first. Similarly, if the justification of an occupation depended on who was the aggressor, you would have a manipulation of charges against both sides.

Add to this mix another consideration. Occupation could be regarded as a substitute for killing the aggressor and thus eliminating the threat. Although in domestic law the defender Dan is permitted to kill the aggressor Alice when it is necessary to do so, recall that a defender nation may not eliminate the aggressor's existence as a collective entity. The defending army may kill soldiers and even capture whole armies, but it may not target the invading state as a whole. There was no way, short of genocide, that the Allies could have eliminated the German or Japanese nations, or that the Israelis can blot out the Palestinians; therefore, there must be some remedy, short of the "homicide" of nations, to achieve security against hostile neighbors. The answer is occupation, the least forcible means available to protect the basic interests of the defender in its long-range territorial integrity and political independence.

Let us look at this phenomenon from the point of view of an occupied state. They have aggressed against another state and thus violated our combined norm 2(4)/51 as stated above, and so their intended victims are entitled to occupy them to ensure their own territorial integrity and political independence. But what about their independence and self-determination? How do we explain their subordination? So far as we accept the legitimacy of defensive occupation, it must be the case that they have *forfeited* their independence, at least temporarily. They can regain their right to political autonomy by engaging in a process of negotiation with the occupying state, convincing them of their peaceful intentions, and entering into a peace treaty.

This way of thinking generates a theory of legitimate occupation. The occupying force must have conditions for ceasing the occupation. The conditions must be related to their interests in peace and security. Israel might properly insist that the Palestinians commit themselves to peaceful relations with Israel and to controlling terrorist incursions against Israeli civilians. But it would be a stretch for the United States to remain in Iraq until a democracy was firmly established unless they could defend the democratic goal in the interests of peace and security in the Middle East.

That we can even pose this novel question about the legitimacy of occupation illustrates, once more, the value of systematically comparing international and domestic law. A theory of partial forfeiture of one's rights—leading to temporary occupation—has some explanatory value in the context of international armed conflict, but it has little explanatory power in the domestic theory of self-defense. To hold this theory, as some philosophers do, you have to explain how aggressors reacquire their rights as soon as the attack has terminated, regardless of the outcome.[19] So the legal right is forfeited, and then the forfeiture is itself revoked, thus bringing the participants back to square one. To speak of legal interests that die and are reborn so quickly strains the imagination. Even more embarrassingly, the argument of forfeiture suggests that the aggressor may be killed with impunity, whether those seeking to kill him know that he is an aggressor or not. This comes remarkably close to treating the aggressor as an outlaw.

These objections against the forfeiture theory are less compelling in an international context. The rights of occupied states are restored by negotiation, and therefore the notion of a revoked forfeiture and reinstatement of rights is not as elusive as it is in the give-and-take of fights between two individuals. Also, if a third party sought to intervene in an occupation, this action would encroach upon the interests of the occupying power and generate a new armed conflict. Whereas in domestic law third parties could conceivably each seek to kill an outlaw without interfering with each other, this is not possible internationally once the defensive use of force has matured into occupation.

Occupation is far better than homicide. Though individual killing is routine in domestic self-defense, entire nations may not seek to eliminate each other in warfare; they may not engage in genocide (killing the nation) or aggrandize themselves by incorporating the occupied land into their own territory (i.e., killing the political entity). Kant correctly rejects the acquisition of colonies or territory by conquest because the permanent subordination of defeated peoples would imply a punitive war rather than a defensive one.[20]

It is important to see that even if a country occupies another as an act of self-defense, defensive occupation is not a form of punishment for the initial aggression. As we noted earlier, Kant's position on punishment follows from his rigorous commitment to the moral equality of all states. Punishment presupposes authority, which must be based, in turn, on a privileged position from which to judge others' wrongdoing. States are not entitled to punish each other, but they may certainly defend themselves, even to the point of temporary occupation.

This inquiry illustrates a fundamental problem in the relationship between the interest protected by the law of self-defense and the nature of aggression. The international community has had a notoriously difficult time formulating a systematic definition of aggression,[21] as well as addressing the related problems of understanding the unlawful use of force or the attack required to trigger a right of self-defense. On a case-by-case, common law basis, criminal lawyers have fared better than international lawyers in trying to define aggression. The common law decentralizes the problem by approaching defensive force as a set of distinct claims based on protecting oneself, other persons, personal property, land, or dwellings. Interfering with these basic interests is the essence of an attack, though of course there is always the problem of determining when such interference is sufficiently serious to warrant a defensive response. Is exhaling in someone's face an attack against his person? Is repeatedly honking your car horn at a pedestrian an attack against him?

These borderline violations of the rights of others are addressed in a body of tort law called *common law of trespass*. How much encroachment on the rights of others is required to constitute a trespass to a person, or to land?[22] In this context, the phrase "borderline violations" suggests a relationship between tort law and international law. Indeed, many tort theorists speak of trespass as a "boundary crossing."[23] The image of neighboring states informs our understanding of individuals as distinct spheres of autonomy. For the purpose of protecting individual liberty, it is correct to say that every person is an island unto himself.[24] Yet determining when a boundary crossing occurs is, in both domestic and international law, a matter of judgment and evaluation. A physical intrusion always constitutes a boundary crossing; an old case in the common law held that cutting a thornbush and allowing the thorns to fall on a neighbor's land constituted a trespass.[25] But interference by sound and light waves does not constitute a physical crossing of the line. (They do not meet the threshold of trespass; however, they might nonetheless constitute a nuisance.)[26] The question is always whether the level of intrusion constitutes a harm that might trigger a right to use defensive force in resistance.[27]

This problem is even more subtle in the Continental legal tradition, because legitimate or necessary force is permissible to protect the autonomy (defined by the set of personal rights) of every person. German courts have agonized in detail over whether specific interests such as privacy and reputation are covered by the general law of necessary defense,[28] whether the specific interest is a right worthy of protection by the use of force. In one case, a German court recognized the right of self-defense to protect the temporary occupation of a parking space (manifested by the parker's wife standing in the spot) against someone else who was trying to take it.[29] No American lawyer would think of justifying the use of force in this situation. The interest being protected would fall between the cracks—not a dwelling, not personal property—not subject to classification under any conventional category for the legitimate use of force.[30]

In the common law way of thinking, there are always two questions: Is the interest covered by the law of self-defense? Is the intrusion sufficiently serious to warrant a defensive response? In the Continental tradition, the first question is submerged in the assumption of a comprehensive interest in personal autonomy. Any interference with autonomy is an attack, but to know whether the person's autonomy justifies defensive force against certain forms of intrusion, you have to decide whether the intrusions constitute a sufficiently serious attack to warrant resistance. The two questions are interdependent; the question of attack leads you to think about autonomy, and the latter leads you back to ponder the former. Concluding that playing loud music in the ear of another is not an attack is equivalent to a ruling that individual autonomy does not include the right not to have loud music played in one's ear. The two concepts—autonomy and attack—might diverge in some contexts,[31] but a good starting point for thinking about them is their reciprocal dependence.

There is no better example of this complex interdependence of ideas than the dispute between the United States and Nicaragua over U.S. support for the contras, a paramilitary opposition group working against the established Sandinista government of Nicaragua. In a hearing before the International Court of Justice, the United States had two lines of argument.[32] First, it claimed that its financial and logistical support for the contras did not amount to a prohibited use of force under Article 2(4). However, if the court found that the U.S. activities did amount to an "attack," the United States then sought to justify its intrusion by invoking the principle of collective self-defense to protect Costa Rica, Honduras, and El Salvador against "armed attacks" by Nicaragua. In both cases— that is, Nicaragua against its neighbors and the United States against Nicaragua—the method of interference was supplying arms and financial support, and there was controversy about whether this level of interference constituted an armed attack. In addition, there was a question as to whether the actions of distinct paramilitary organizations could be attributed or imputed to the states involved. Under the conventional way of thinking about international law, there would be no violation unless these actions were deemed to be the actions of the state.

The first argument raises an unavoidable problem, because some forms of getting involved in the affairs of others are permissible. For example, there is nothing wrong with publishing articles in the *New York Times* condemning the Sandinista government or even with broadcasting propaganda messages to encourage the local population to rise up and overthrow the existing regime. Providing financial support to the opposition might be permissible, but there must come a point where encouraging regime change becomes an unlawful intrusion in the affairs of another state. The court drew the sensible distinction between supplying funds to the contras and "arming and training" them.[33] Generating armed resistance amounts to armed intervention. The United States tried to advance the policy of legitimate regime change, but apparently this argument did not generate a sufficient legal basis for deviating from a strict principle of sovereign independence, regardless of the moral quality or political respectability of the regime.

Having concluded that the United States improperly intervened in the affairs of Nicaragua by arming the contras, the court had to confront the issue of whether this action was justified as collective self-defense on behalf of Costa Rica, Honduras, and El Salvador.[34] The case might have qualified as collective self-defense had the Sandinista government actually been using force against those nations, in violation of Article 2(4). The court concluded that the Sandinistas had not engaged in an armed attack against their neighbors, but nonetheless evaluated the U.S. claim of collective self-defense. The court reached a number of conclusions that are very much at odds with our claims about légitime défense in the name of the international community.

We should note that in the French domestic law of legitimate defensive force (légitime défense), the law of self-defense extends to third parties under attack. There is no need for the consent of the third party. This is the case under most Continental European legal systems, which equate the right to defend oneself with the right to defend others, and treat them under a single statutory provision. In the United States and other common law systems, though, the law distinguishes between self-defense and defense of others and treats them with different provisions. For example, the Model Penal Code has a separate provision for "defense of other persons" that says nothing about the *consent* of the other.[35] This issue is the sole focus of chapter 3. The same principle holds in all modern legal systems: the victim need not consent to be rescued. There is not much discussion of this point in the contemporary literature. It is taken for granted that because the interveners seek to defend the legal order—to defend the law itself—they need not have the particular consent of the victim.

Nonetheless, the International Court of Justice came to two conclusions about the conditions for collective self-defense that fly in the face of domestic law. They required the neighboring countries that the United States was seeking to protect to express two separate states of consciousness: first, that they were being attacked, and second, that they desired intervention by the United States.[36] It is unclear if these requirements are appropriate. One can imagine cases where the victim is unaware of the attack (at least in domestic law) and nonetheless needs protection; in the international arena, it might be more difficult to imagine a case of actual attack that the leaders of the victim state would fail to recognize. But suppose the leaders of a nation were all kidnapped and a puppet regime installed. Would the international community still insist on an articulated consciousness of being attacked?

The second conclusion of the International Court of Justice was that the victim state must consent to collective self-defense. There are two possible reasons for this. The court's sense of the international legal order might have been too weak to uphold the argument that, by defending the interests of a third party, the defender was upholding an international legal order comparable to domestic legal orders. The second reason is more pragmatic: a fear of states mixing in the business of other states and exploiting claims of defensive necessity to invade them and control their internal affairs. Interestingly, this pragmatic argument has no parallel in the defense of other persons under domestic law. The defended person has nothing to fear from temporary assistance rendered to fend off aggressors, but defensive force in the international context entails more than fending off an attack. It implies ongoing military action, the stationing of troops and armaments on the territory of the defended state, and the inevitable impact on civilian lives. Understandably, states do not want to be defended by allies who might ultimately interfere in their lives even more than the aggressor threatens to.

B. Self-Defense as a Right

We conventionally refer to self-defense as a right. As we discuss in greater detail in the next chapter, the UN Charter uses the term "inherent right" in English to describe self-defense, but the phrase "natural right" in French. We agree on the word, but what does it mean in this context? "Rights" bears the connotation of being *in line* with the law. Indeed, virtually all European languages use the same word for both personal rights and the law itself. Thus, Germans must distinguish objective Right (*objektive Recht*) and subjective or personal rights (*subjektive Rechte*). In French, *le droit* refers both to law as in "rule of law" (*la règle de droit*), as well as a "right" to do something. Acting with a right, in this dual sense, brings one into conformity with the legal order. Justification has the same connotation. *Jus*, after all, refers to law in the same elevated sense as we find in the European words for Right (e.g., *Recht, droit, derecho, prava, jog*).[37] Justification means to render right (*jus-facere*).[38] The action rendered "right" is the exercise of force in self-defense—force that would otherwise constitute a crime in its own right. It is right because it stands in a certain relationship with the aggression that it seeks to repel.

As we have already noted, the right of self-defense bears no similarity to rights in private law. It is not a right *against* the aggressor, but rather an expression of the Law (the Right) seeking to reestablish itself against external aggression. To put it another way, the defender's lawful autonomy is compromised by the attack, and using defensive force is a way of vindicating his autonomy and reestablishing the supremacy of the law. This is the sense in which the defender has a right to use force—the sense that the Right is on his side.

The claim that the defender has a "right" to *kill* in self-defense is a misleading statement of the legal relationship between aggressor and defender. The defender does not have the right to kill in the same sense that he has the right to demand the performance of a contract. When you have a right to do something, you are entitled to enjoy exercising that right, to do it as an end in itself, but if the defender "enjoys" the use of force too much, his legal position is undermined. For example, when Bernard Goetz shot the four black youths on the subway, many charged him with a motivation of racial animus rather than accept his claim of defensive necessity. Insofar as his motives were racist, his position was much less believable. He could not get up in court and say, "I had a right to kill, and I enjoyed it." Nonetheless, many gun advocates celebrate the expression "Make my day" as a motif of self-defense.[39] In other words, if you attack me, then you make my day by giving me the right to use deadly force in response.

The idea that an aggressor vests a right to kill in response overstates the nature of the right to respond to aggression. The purpose of the response must always be to repel the attack. As we pointed out earlier, the response is described in terms different from those that apply to an act that infringes on the norm defining the crime. Whereas the aggressor may intend to kill, the defensive reaction is described not as an attempt to kill, but only to ward off the aggressive attack. This restatement, conceptually linking defensive force with avoiding the attack, is made explicit in the terminology of the leading criminal codes in the world. The Model Penal Code requires that the use of force be "for the purpose of protecting [oneself] against the use of unlawful force." The German Penal Code stipulates that the purpose be "to avert the attack."[40] The Rome Statute defines self-defense for killings that would otherwise be war crimes as force directed "against an imminent and unlawful use of force."[41] It is safe to generalize and conclude that no criminal law in the world deviates from this pattern. It would be a great surprise if somewhere a criminal code read, "And if someone attacks you, you have the right to kill him."[42]

The right of self-defense, in the strong sense, implies both the defender's right to avert the attack and the aggressor's lack of a corresponding right to use force to "defend" himself from the defense. The latter point is expressed by the thesis of incompatible rights. If a mother has the right to abort a fetus in the first trimester, for example, no one has the right to defend the interests of the fetus. That is, one party's right to act implies a corresponding denial of the right of those who are affected to resist the rightful use of force. Rights in general, and justificatory claims in particular, have a subordinating effect; for the purposes of exercising his right, at least, the right holder dominates those who may have inconsistent interests.[43]

The same principle seems to apply in international law: if the United Nations authorizes military intervention under Chapter VII of the Charter, this right of intervention presumably excludes the right of self-defense for the state that is subject to the UN's military measures. But other forms of international intervention are more dubious and controversial. For example, if Serbia could have mounted an effective resistance to the invading NATO forces, it would arguably have had the right to do so. The NATO force was acting outside of Chapter VII's system of authorized force. NATO might have made an argument based on collective self-defense of the Albanian national interest in Kosovo, but this argument is controverted for several reasons. For example, there is some doubt about whether self-defense could operate on behalf of the Albanian nation rather than an independent state. Article 51 does, after all, refer to the member states of the United Nations. Yet it is also true that nonstate national entities have received recognition in the period since the adoption of the Charter, notably in the Genocide Convention of 1948 and the International Covenant of Civil and Political Rights adopted by the General Assembly in 1976.[44]

"Strong" rights, then, exclude the possibility of resistance, but rights are not always applied in their strong sense. "Rights" may also refer to permissions that are indeed compatible with permissions to resist. Kent Greenawalt seems to subscribe to this version of rights when he argues that both sides in a conflict might be warranted in using defensive force.[45] In referring to these rights as permissions, we have something much less august in mind than acting in such a way as to render the action lawful.

Sometimes having a right to an activity simply means that no governmental agency has the authority to prohibit the activity. This is what people mean when they refer to a "right to free speech." In the United States, no state or federal authority can censor or prohibit speech on the basis of the message conveyed.[46] In this limited sense we have a "right" to write this book. Rights as "permissions" or "warrants" tell us only that the rights bearer will not be prosecuted for exercising the right. But we learn virtually nothing about the implications of rights for legal relationships with others.

It is difficult to decide whether the strong or weak sense of a right to self-defense should prevail. Current usage seems to support the strong version of rights; the inherent right of self-defense implies that the aggressor is not permitted to respond to defensive force with the use of force. The doctrine of self-defense speaks to the relationship between states caught in an armed conflict, and its purpose is to distinguish neatly between the lawful and the unlawful, between aggressor and rightful defenders. But if the notion of rights is reduced to a set of permissions, these sharp profiles lose their definition. In the international arena, with its deficit of courts and other enforcement mechanisms, the prospect of prosecution pales in importance. In the diplomatic and legal interaction of states, it is important to be able to make well-defined charges of aggression and to respond with well-grounded claims of justification. The proper concept for our purposes, then, is the strong sense of rights.

C. Self-Defense as a Duty

Referring to the right of self-defense as "inherent" or "natural" is a good way to sidestep arguments about its foundations. After all, where do rights come from? How do we argue for their existence? One view, explored in the strong theory of rights, is that they derive from the law. They represent the ability of each citizen to vindicate his or her autonomy under the law. In the Kantian system of law, all of these issues hang together. The notion of law, or Right, is based on the maximum possible external freedom for all.[47] Aggression compromises this freedom, and therefore the defender is entitled to vindicate his freedom by repelling the aggressor.

An entirely different methodology for deriving rights is grounded in the recognition of preliminary duties. For example, the Belgian defense of humanitarian intervention is based on a simple syllogism: Every state has a duty to intervene to prevent human disasters, and therefore every state must have a right to do so.[48] This is an unusual way to derive a right of military intervention, one that is generally foreign to Western legal culture. Yet there is one ancient legal system that does derive its principles of defensive force from a duty of rescue, and therefore we should take a closer look at that system. There is much to learn from the ancient sources, in this case Jewish law, based on the Bible and the Talmud.

Jewish law is based, ultimately, on the narratives and laws recited in the first five books of Moses. The ancient Hebrews then developed an oral law as a companion to the Bible, which they passed down, by memory and rote, from generation to generation. In the third century of the Common Era they finally put the oral tradition to paper and produced the first code of Jewish law, called the Mishna. The rabbis then began an intensive period of debate and commentary about the Mishna and its implications, generating an extraordinarily nuanced body of legal literature called the Gemara. The Mishna and the Gemara, plus the early commentaries on the Gemara, constitute the revered, multivolume book of study known as the Talmud.

In a very thoughtful and respectful chapter on the Talmud, the Canadian legal comparativist H. Patrick Glenn quips that one might consider spending a year, perhaps even "twenty or thirty years at it, or more," studying the multivolume work.[49] Many scholars do in fact devote their lives to Talmudic study as an act of religious devotion.[50] Our comments in this section are based on much less exposure. Fortunately, there are only a few pages of the Talmud devoted to the subject of self-defense, and they form the foundation of our comments on self-defense as a duty.

The root of this Talmudic discussion is a biblical passage, Leviticus 19:16, which commands, in Hebrew, *Lo taamod al dam reacha*, literally, "Do not stand on thy neighbor's blood." Making sense of these cryptic words has long been a challenge to translators in all languages. Luther rendered the phrase as *Du sollst auch nicht auftreten gegen deines Nächsten Leben*, which means basically, "You should not rise up against your neighbor's life." Luther's point is obscure, but the gist of his translation seems to be shared by the new French Catholic version: *Et tu ne mettras pas en cause le sang du prochain* ("Do not put into question the blood of thy neighbor"). But these translations miss the idea, accepted in the Talmudic discussion, that standing on one's neighbor's blood means standing and doing nothing, or standing idly by while one's neighbor suffers. The Talmud interprets Leviticus 19:16 as implying that one is under a duty to rescue those in distress. Our preferred translation would be "Do not stand idly by while your neighbor bleeds."

The legal duty to rescue those in danger or injured by an accident is well accepted in Continental European jurisprudence. Anglo-American common law has always been more skeptical of this duty. There might be a *moral* duty to rescue, but this moral duty is not recognized as a legal obligation. In special cases, however, where there is a personal duty to the person in distress—as in the duty of a mother to her child—the failure to rescue might generate liability for the ensuing harm to the endangered person.

In the Talmudic view of the world, there are distinct spheres of possible defensive action. The first is the private realm (where every man's home is his castle, his own private island); the second is the public sphere, where people mix and jostle on the street. For defensive action in the private sphere we have a specific biblical license, the passage in the Book of

Exodus, not long after the Ten Commandments, that provides the basis for forcible self-defense in cases of burglary: "If a thief is found breaking in, and he is struck so that he dies, there shall no blood be shed for him" (Exodus 22:1). There shall be no "shedding blood" (i.e., punishment) for killing a thief caught in the act of breaking in. In other words, killing the burglar is justified. The Talmud goes on to explain why deadly force is permissible even when it appears that only property interests are at stake: the homeowner is likely to resist the burglar, which will escalate the conflict and lead the burglar to threaten the life of the homeowner. So ultimately, the use of deadly force is justified as the protection of life.[51]

Defending one's home is a relatively easy case. The intruder crosses a clearly demarcated boundary by trespassing against a zone of intimacy that is fundamental in any legal system that takes individuals and their possessions seriously. When one person breaks into another's private dwelling, there is no doubt, to borrow a phrase from a neighboring passage in Exodus (21:28), about whose ox is goring whose.

When the interaction of two people in the public sphere degenerates into a fight, however, one can never be entirely sure who started it. True, someone swings first, but behind the swing may lie prior verbal and other potentially provocative behaviors that can make it difficult to distinguish between a chance melée and an unequivocal attack. It makes sense to think about imposing a duty to retreat from these potentially ambiguous fights in the public arena, but no one has ever seriously suggested that a homeowner has a similar duty to retreat from his own home rather than use deadly force against an intruder. This explains why the rabbis did not try to extend the model of the private home into the more difficult sphere of fights breaking out in the public arena. And there is no specific textual base for formulating a right to use defensive force against attacks that occur in public.

Though the Bible is silent about a general right to use defensive force, the oral law as codified in the Mishna recognizes the right to kill an aggressor in certain specified cases. The Hebrew term *rodef*, usually translated as "pursuer" or "aggressor," provides the linchpin for both a duty and a right to use deadly force. The text of the Mishna specifies three cases in which one should kill a rodef: to stop him from chasing another "in order to kill him," to stop him from chasing a man, and to stop him from chasing a betrothed woman. The last two cases do not specify the purpose of the rodef in pursuing his intended victim, but the context makes it clear that the purpose is illicit sex. Unfortunately, what makes the sex illicit in these two cases is not so obvious. Most people think the aggressor must intend to use force against the man or the betrothed woman and thus be guilty of homosexual or heterosexual rape. (But this is not clear. The prohibited nature of the sexual liaisons might be enough to justify preventing the aggressor from contaminating himself as well as the other party.)

The Mishna provides some guidance by also delineating three cases in which deadly force is not permissible, even though the rodef may be threatening to commit a capital crime. If the pursuer is after an animal with the intent to have sex with it, one may not intervene. Although the completed crime was subject to the death penalty, it did not follow that a bystander could intervene and kill the offender before the act. This same conclusion is reached where the pursuer and probable offender threatens to desecrate the Sabbath or engage in idolatry. All three of these acts represent capital offenses in Talmudic law, but none of them directly endangers the welfare of a human being. Therefore, bystanders are not entitled to use deadly force to prevent the crimes from occurring. Upholding God's law, even in matters as basic as sexuality, the Sabbath, and proper worship, is still not enough to justify deadly force; another person's life or sexual purity must be at stake. If Talmudic thinking were based exclusively on a principle of protecting the sacred order, one might expect to find a broader right to intervene; there would be no reason why individuals would not be empowered to prevent bestiality, desecration of the Sabbath, or idolatry.[52]

Oral law and the Mishna clearly go beyond the biblical prescriptions. The term rodef, in the sense of "aggressor," does not even appear in the Scriptures. Yet the rabbis of the Talmud found warrant for their doctrine in Leviticus 19:16: "Do not stand idly by while your neighbor bleeds." It does not matter whether the danger derives from human or natural sources. The attack by an aggressor is treated as analogous to a natural disaster that befalls the potential victim, and intervening to stop the aggression is like inserting oneself between a falling rock and the innocent person below it.

Grounding self-defense in a duty to rescue implies, paradoxically, that the defense of others is more easily justified than the defense of self. Western legal systems typically take the defense of self to be the basic principle, and the defense of others is seen as an extension of the right to protect oneself. This process of generalization follows, as we have seen, from the idea of defending the legal order against the aggressor, and a generalized principle of justification treats the defense of others exactly the same as the defense of self, which is implicit in the Continental doctrines of necessary or legitimate defense.

Treating defensive force as an act of rescue transposes this relationship between the self and the other. In the Talmudic approach, the defense of others is primary and the defense of self derivative. But making the leap from rescuing others to defending oneself is not so easy. One could claim that everyone has a duty to preserve himself that is at least as strong as his duty to rescue others; everyone is his own closest "neighbor," and therefore no one should sit idly by if he himself, as well as his neighbor, is bleeding to death. This seems a bit strained, though, for the biblical passage appears to teach altruism more than self-love. Perhaps we simply need to be reminded of our duty to rescue others. There is apparently no corresponding need to issue an imperative to save ourselves.

In another Talmudic passage, devoted to the implications of the home-owner's right to kill an intruder, the rabbis assert, "The Torah holds, 'If someone comes to kill you, rise up and kill him first.'"[53] The passage suggests that Jewish law recognizes a general right to defend oneself on the street, as well as within the confines of one's own home, but the reference to the Torah is used here in a wonderfully ambiguous way. Narrowly defined, the Torah refers to the five Books of Moses (Genesis, Exodus, Leviticus, Numbers, and Deuteronomy), but as we have suggested, nowhere do these books explicitly recognize a general right of self-defense—only the justification to kill thieves in the act of breaking in. As used in the Talmudic recognition of self-defense, "Torah" refers not to the Bible, but to the entire body of Jewish law, including the unwritten principles of the oral tradition.

We speak here of a right of self-defense, but we should not forget that the basic Talmudic idea is that of duty, in particular a duty to God to fulfill the divine commandments. The notion of individual rights in the modern sense does not appear in either the Bible or the Talmud. This may explain why the focus is on the duty to rescue others. Rescuing oneself is an expression of the instinct to survive and does not require legal recognition of a duty to act. Even within the framework of a religious legal system based on a single text, the lawyers (or in this case, the rabbis) engaged in interpreting the text are permitted to assume that some matters, such as the right to defend oneself, are so obvious that the lawgiver need not expressly mention the right in laying down the law.

A duty to rescue a person in danger would presumably entail a right to act to fulfill that duty. The assumption is not only that "ought" implies "can"—the principle often associated with Kant—but that "ought" implies "may." Grounding the right of self-defense in a duty to rescue implies, however, that the notions of right and duty are coextensive. But they are not. There are clearly instances when we have the right to act but are not *obligated* to act. We may use our discretion to refrain from exercising our rights. Our foundational idea is the individual's right to protect himself and vindicate his autonomy, so the legal system should encourage a notion of individual rights that is broader than the duty to act on that right. Indeed, this is the position that has evolved in Western legal systems. Whether one acts to protect another's body or property is left to individual judgment. Rights to use force then come into focus as permissive, or optional, rather than mandatory; extralegal considerations such as compassion or fear might lead one to refrain from the use of permissible force. In the international sphere, the restraint is often prudence. It might have been legitimate for the West to intervene in Budapest in 1956 or Prague in 1968, but legitimate concerns about war with the Soviet Union instilled a calculated sense of restraint.

Considering defensive intervention as a duty-based act of rescue seems to threaten the universality of the right to use defensive force. After all, could one be under a duty to rescue everyone? In modern legal systems, the scope of the duty to rescue is limited to family members and a set of persons with whom one has a special relationship of trust and responsibility. If we begin from this premise of a limited relational duty, the right to use defensive force would be similarly constrained by particular relationships. The biblical duty not to stand idly by while one's neighbor bleeds appears to extend broadly to all members of the community, which in this context means all Jews. The Talmud suggests that the scope of this duty stops short of idolaters.[54] Though modern thinking about self-defense is pitched to multiethnic, diversified societies, the Talmud approaches the duty to rescue (and therefore defensive intervention) in the context of a closely knit, homogeneous community. So far as the duty to rescue is limited by communitarian empathy, so, too, will the right of self-defense be circumscribed and available only to protect persons with whom one stands in a close relationship.[55]

Grounding defensive force in the duty to rescue also expresses the communitarian orientation of Talmudic jurisprudence: all Jews are responsible to each other for each other's sins on the Day of Atonement and for each other's welfare in daily life. Only hyperbole can capture the Jewish passion for interdependence: "If one saves the life of [a Jew], the Scriptures treat him as though he has saved an entire world."[56] This statement, found in the Mishna among legalistic propositions about interrogating witnesses, follows an analogous claim about the implications of killing a single Jew: it is as though one has killed an entire world. This poetic testimonial to the value of human life derives in part from the mystical Jewish view that each person carries within him not just the seed of his offspring but the seed of infinite generations yet to be born, and these unborn generations are sacrificed by every act of killing—or of failing to rescue.[57]

As a general matter, the Talmudic conception of self-defense has considerable difficulty coping with the concept of permissive intervention. The notion of permissive action—where one has a choice whether or not to intervene—presupposes a concept of rightful (but not duty-bound) action. Permissive action inhabits the middle ground between the range of individual rights and the onset of duty, but making the notion of duty primary, in Talmudic fashion, squeezes out the central position occupied by permissive action.

Yet Rashi, a medieval commentator on the Talmud, suggests that a homeowner's killing a burglar is permissive, rather than mandatory.[58] Unfortunately, to make this point he must treat the killing as a trivial act, and how he achieves this requires a word of clarification. To express the notion of justifiable conduct, both the Bible and the Talmud use the suggestive expression "He has no blood."[59] This phrase is usually taken to mean that there is no blood guilt for killing him—no blood guilt because the killing is considered proper under the circumstances. Rashi turns this expression on its head and claims that "He has no blood" means that as a result of breaking in, the burglar must be treated as equivalent to someone without blood—as a dead man walking.

This is a version of the familiar theory of forfeiture as a rationale for self-defense. The burglar forfeits his life by breaking in, and thus the homeowner may kill him if he wishes, for there is no legal prohibition against killing a dead person. This way of thinking about self-defense raises numerous problems, not the least of which is explaining why the forfeiture is revoked and the burglar's life reinstated as soon as he flees from the house. Neither the Talmud nor modern secular legal systems recognize the right of the homeowner who has been the victim of a burglary to hunt down the burglar wherever he is and kill him, and yet this right would seem to follow if the burglar had indeed forfeited his life.

Recall Gardner's "remainder" thesis, the modern conception of self-defense as justification that presupposes that killing the aggressor leaves the shadow of a wrong even when the conduct is justified. Looking at such a killing as the fulfillment of a legal duty cuts against this perception. It stops us from appreciating that thwarting aggression is an invasion (albeit a justified one) of human life that should be taken only as a last resort.

In the end, it appears that a number of significant and intuitive features of self-defense are sacrificed when we try to derive the right of self-defense from a duty to rescue. We lose the sense that the right to rescue others is universally attached to everyone, regardless of whether or not the party deserves to be rescued. The notion of self-defense as permissive intervention—an act that may be exercised or not—falls victim to the duty to intervene, an obligation in place of an act of compassion. Further, deriving the right from the duty undermines the nuanced conclusion of the remainder thesis: that using deadly force to avert an act is justifiable, but nonetheless a breach of the victim's right to life.

The point of this discussion has been to explore the possibilities suggested by the Belgian case for forcible intervention on the basis of a universal duty to aid others in distress. Their universal modern argument has ancient roots, not surprising in light of the general preference for duties over rights in ancient legal cultures.[60] On the whole, however, the evolution from a culture of duty to a law based on rights is a worthy transition. The Talmudic alternative has its attractions but, on balance, it should lead to enhanced appreciation of the rights-based model of self-defense that has come to dominate the mind of Western legal thinkers.

2. FOUR MODELS OF SELF-DEFENSE

It is one thing to know the doctrinal foundations of self-defense; it is quite another to see them in practice. In this context as well as others, the elements of the law take on differing connotations depending on the underlying philosophical conception of self-defense. In fact, four distinct theories of defensive force shape the way we interpret the substantive law, whether we are discussing a shooting in a graffiti-ridden New York subway car or launching missiles on the border between India and Pakistan. By examining these conflicting models and value systems, we can begin to see the larger issues at stake in every dispute about defensive force.

A. Self-Defense as Punishment

Our commitment to justice and to the symbolic expiation of evil often leads us to consider self-defense as a form of just punishment: the individual acts in place of the state in inflicting on wrongdoers their just deserts. If Troy Canty, Barry Allen, James Ramseur, and Darrell Cabey were in fact muggers, then this rough principle of justice holds that "they got what they deserved" from Bernard Goetz. These are the exact words of a black witness, Andrea Reid. Present with her baby in the subway car at the time of the shooting, she was also afraid of the four "punks who were bothering the white man." Goetz's defense attorney, Barry Slotnick, referred to her words "They got what they deserved" dozens of times in the course of the trial. Sometimes he paraphrased the comment in a more respectable legal idiom: "They got what the law allowed."

Goetz became a folk hero because, as many people saw it, he brought the arrogant predators to their knees. Yet even under the most punitive theory of giving criminals "what they deserve," there remain questions of fact. Did these kids have records long enough to support the judgment that they were criminals and predators? Or is the public perception of Canty, Cabey, Ramseur, and Allen as criminal types largely a function of their race and youth? When our passions seek gratification—when our lust to avenge evil gains the upper hand—we don't always ponder the facts and weigh the gradations of evil, or their fitting punishment.

The idea that people should be rewarded and punished on the basis of their character and their lifelong behavior may express a principle of justice, but it is a principle better suited to infallible divine punishment than the imperfect human institutions of the law. In fairy tales, the witch may receive her comeuppance in the end, but surely it is not the business of human institutions—not to mention a lone rider on the subway—to determine who is a witch, or a wicked person, or an unreformable criminal. The law wisely limits itself to the question of whether a particular act constitutes a crime that merits punishment, or whether, in the context of self-defense, a particular aggressive attack properly triggers a defensive response. The general character of the suspect is important neither for human punishment nor for the assessment of whether defensive force was permissible in a particular situation. Some people who passionately sided with Goetz's victims may have thought that when Barry Slotnick was assaulted a few weeks after the trial, he, too, got what he deserved. They are entitled to their opinion, but their passion for justice on the streets should not be heard in court. Nor should Slotnick's repeated reiteration of Andrea Reid's words "They got what they deserved" have been considered a persuasive argument for the proper scope of self-defense.

In the international arena, the urge to punish often overwhelms proper analysis as well. This certainly appeared to be true in the context of the U.S. invasion of Iraq in 2003 when U.S. policymakers tried to stretch self-defense to fit the facts while pointing incessantly to the great evil represented by Saddam Hussein. His crimes against the Kurds and other political opponents were recited in detail. Whether the purpose of regime change was punitive or philanthropic was never clear, and the two became readily intertwined with our government's claim of self-defense against the prospect of weapons of mass destruction.

As we noted earlier, Kant maintained that punitive wars were a logical impossibility. Punishment, he claimed, presupposes superior authority, as parents are in a position to punish their children and God can punish human beings, but because, in Kant's view, all states are moral equals, no state has the authority to punish another. Although it is true that punishment might be imposed by the world community, in the same sense that the community imposes punishment at the domestic level, this is both rare and insufficiently theorized. The power of international institutions that purport to punish in the name of the world community, such as the UN Security Council, are invoked in only a small number of today's conflicts. Reprisals (or putative reprisals, in which one state thinks it has the authority to punish another) are much more commonplace in international affairs.[61] For good or ill, superpowers think they are in a position to discipline their wayward adversaries.

In the final analysis, we have to keep in mind that self-defense is not punishment. The purpose of a defensive act is not to inflict harm according to the desert of the aggressor; its purpose is to repel the attack. Working out this issue in the Middle Ages was an important step forward in understanding the use of force. As motives become more complex and confused, as emotions cloud the issues, individuals and states sometimes forget this simple distinction. Although we might feel that an aggressor gets his just deserts when he is killed in self-defense, it is our protection against his attack—not his punishment—that justifies killing him.

B. Self-Defense as an Excuse

An alternative approach to self-defense shifts the focus away from a passion to punish wrongdoing and directs it instead to the personal plight of the defender. In the closing portions of his summation to the jury in the *Goetz* case, Barry Slotnick played on this theme. He stressed Bernard's fear: his back against the steel wall of the subway car, with no choice but to strike back. Fear invokes the primordial form of self-defense in English law; from roughly the thirteenth to the sixteenth century, the plea of self-defense, called *se defendendo*, applied whenever a fight broke out and one party retreated as far as he could go before resorting to defensive force. His back had to be, literally, against the wall.

If he then killed the aggressor, se defendendo had the effect of saving the defendant from execution, but it did not negate other stigmatizing effects of the criminal law. The defendant forfeited his goods as a token sanction for his having taken a human life. The murder weapon was also forfeited to the crown as a *deodand*, or tainted object. Killing se defendendo was called *excusable homicide*, for though the wrong of homicide had occurred, the circumstances generated a personal excuse that saved the killer from execution. The defense of se defendendo springs more from compassion for the predicament of the trapped defender than from a commitment to justice. If it is likely that we would all act the same way in the same circumstances, we can hardly condemn and execute the defendant who had no choice.

Although this theory of excuses plays a large role in domestic criminal law, it seems to hold less sway in international law, because all actions between states are assumed to be voluntary. States do not go insane; they cannot excuse aggression because their people are hungry, because their leader makes a mistake about their legal obligations, or because they feel in danger of some future attack. There is some recognition of duress in international law, not as an excuse but as a factor reducing the wrongdoing of the state in suits against the state for compensation.[62] The knottiest problem is whether a reasonable mistake about the intentions of an aggressor should be considered in the analysis of self-defense. It would be difficult to disregard these reasonable mistakes altogether, because defending states must always act on the basis of appearances. (Suppose the Israelis were mistaken about Egyptian intentions in 1967. Would they have lost their apparent right to defend themselves?) We shall return to this issue in due course. It is sufficient at this point to recognize that defensive action by states is treated as a justification rather than an excuse for resisting aggression. Protecting your territory and political independence is the right thing to do; it is not simply an involuntary response for which we ought to have compassion. The general moral question about excuses for the behavior of states must await further reflection. If it turns out that excuses are excluded from relations between states, this might be the most significant feature that distinguishes individual behavior in domestic law from collective action in the international sphere.

C. *Self-Defense as the Justified Defense of Autonomy*

English lawyers of the fifteenth and sixteenth centuries paid close attention to the jurisprudence of the Bible, and in Exodus 22 we read that there is no blood guilt, no "taint," in killing a thief who seeks to break into one's home at night. This way of thinking was at odds with the early common law, which approached the subject of self-defense solely as a matter of excuse and did not recognize a justification that would totally exempt a "manslayer" from liability. The minimal penalty for those who claimed se defendendo was always the forfeiture of goods.

Eventually the British Parliament closed this gap in the common law, enacting a statute in 1532 that essentially licensed the killing of robbers and other assailants on the public highway. This statutory defense came to be called *justifiable*—as opposed to excusable—*homicide*. The defense is not based on compassion for someone with his back to the wall; rather, it expresses a right to hold one's ground against wrongful aggressors. The defendant's claim is not "I could not do otherwise" but "Don't tread on me!"[63] As the leading legal scholar of the seventeenth century, Sir Edward Coke, said of this defense, "A man shall never give way to a thief, etc. neither shall he forfeit any thing."[64] The result of justifiable homicide, as the 1532 statute prescribes, is total acquittal without any forfeiture of goods.

This version of justifiable self-defense is appropriately called "individualist," for it is guided by the imperative to vindicate individual autonomy; its philosophical champions are John Locke and Immanuel Kant. Kant conceived of an unqualified right of self-defense as the foundation of a liberal legal system, in which each citizen recognizes and wills maximum freedom for himself as well as for his fellow citizens. That the legal system should be organized on the basis of maximum freedom was, for Kant, the implication of pure reason in human affairs. Relying implicitly on this tradition, Slotnick invoked the individualist theory in his opening statement on behalf of Bernard Goetz:

> No one can ever take away your inalienable right to protect your property or your life or your family. No one can walk up to me and say, "Give me that watch," "Give me your ring," "Give me five dollars." And if they do, heaven help them if I'm armed, because I know what the law allows.

This vision of autonomy depends on an analogy between persons and states. The individual is an island, self-reflexively sovereign, as nations are sovereign over their territory. Indeed, the individual right of self-defense makes sense as an extension of the idea that nations can use force to maintain dominance over their own people and their own territory. Because self-defense is so strongly connected with what it means to be a person or a nation, we speak readily of self-defense as a natural right, as in the French version of the UN Charter, or an "inherent right," as in the English version. In this context, the principles of domestic and international law are so tightly interwoven that it is difficult to know which came first.

The individualist theory has always expressed itself most strongly in the protection of one's home. Any intrusion against one's "castle," one's refuge from the heartless world, is intolerable. The individualist theory would vindicate the use of any means necessary to defend one's home against an attempted intrusion. Perhaps, as Kant would say, our moral concern for the welfare of others could lead us to choose not to exercise our right of defense, but liberal writings leave no doubt that freedom entails the option to resist all forms of encroachment.

D.   Self-Defense as the Defense of Society

The individualist view can be contrasted with a "society-based" variation of justifiable self-defense. The difference between the two is expressed in a very loose (as opposed to very strict) approach to proportionality. The extreme version of the individualist defense rejects proportionality altogether. Any encroachment on an individual's rights represents an intolerable violation of personal autonomy, and the affected individual can do everything in his power, deploy all necessary means to end the encroachment and vindicate his autonomy. As we have seen, this is the dominant approach toward using force to preserve the territorial integrity of states.

The society-based variation of justifiable self-defense rejects absolutes like the imperative to secure one's rights, one's autonomy, and one's private physical space, whereas the individualist treats every person as an island, entitled to full sovereignty in his own domain, as though people were states. The individualist approach ignores our interdependence, both in shaping our sense of self and in cooperating in society for mutual advantage. The alternative, society-based view regards the aggressor as another member of the same society of interdependent selves, with interests that cannot be ignored, even if he acts wrongfully in aggressing against someone else. The defender is obligated to consider the aggressor not merely as an intrusive force, but as a fellow human being. In the end, the defender's interests may be worth more, but those of the aggressor are not totally discounted.

If one recognizes the humanity of the aggressor, it follows that in some situations the defender will be required to absorb an encroachment on his autonomy rather than inflict an excessive cost on the aggressor. If the only way to prevent intrusion and nonviolent theft in one's home is to kill the aggressor, the defender may have to forgo defensive force and risk losing his property; he must suffer the minor invasion and hope that the police will recover his goods, since the alternative of killing the aggressor is too costly and callous a disregard of the human interests of the aggressor.

In the 1950s and 1960s, a strong defense of this society-based theory emerged from the seemingly uncontroversial general view that criminal law was meant to further the public good. After all, who could be against the public good? Criminal law reformers claimed that the purpose of punishment was primarily to encourage people to act in a socially desirable way, and the determination of what was socially desirable depended largely on the assessment of the costs and benefits of acting in particular ways.

For a time, viewing self-defense as a measure to further the public welfare led courts and legislatures to eschew all absolutist thinking about the right to defend one's autonomy against aggressive attacks. In cases of burglary, for example, some lawmakers insisted that for a homeowner to be allowed to use deadly force against an intruder, he must fear violence to himself or the other occupants; the fear of theft would not be sufficient to justify fighting off the burglar with force endangering his life. A good illustration is the decision of the California Supreme Court in *People v. Ceballos*, which held that injuring a burglar with a spring gun (an automatic weapon triggered by entry) could not be justified.[65] Because no one was home at the time of the intrusion, the burglary did not subject an occupant to the risk of violence, and therefore the only interest at stake on the side of the homeowner was his property. The court held that defending property alone did not justify the use of deadly force against the burglar.

The social theory of self-defense says, in short, that burglars and muggers also have rights, and the rights of victims must therefore be restricted when their exercise inflicts an excessive cost on those who attack them. It's fair to say that though the public at large may have supported this philosophy a generation ago, their feelings about crime and the rights of criminals have shifted dramatically since then.

The international law of armed conflict, however, has not reached the stage where states have no right to use force to defend property. Very few world leaders questioned the right of the United Kingdom to defend the Falkland Islands against Argentine aggression, regardless of how few British citizens actually lived there. The same principle would hold with regard to totally unoccupied territorial possessions.

Yet some of the principles underlying the social conception of self-defense have managed to penetrate the international law of war, most notably in the evolution of international humanitarian law. This body of law has evolved on the basis of the Geneva Conventions of 1949, which laid down the rules of war and generated protection for prisoners of war and civilians not engaged in combat, rules constituting the field of jus in bello, defining how wars are properly fought. The basic moral point underlying international humanitarian law is that enemy soldiers are also human beings; when they are sick or taken prisoner, they regain the protection owed to all members of the human species. If civilians never don a uniform or carry a rifle, they are entitled to the same basic rights that they have always enjoyed. The fact that their nation has gone to war does not forfeit their inherent right to be treated as human beings.

Just as recognizing the humanity of aggressors imposes restrictions on the exercise of defensive force, so this same moral insight imposes restrictions on the way wars are fought. We must never forget that war, aggression, and self-defense represent conflicts among human beings who—from a moral point of view—are created equal. As Lincoln recalled that famous principle in the midst of the American Civil War, so we should never forget that our enemy is essentially "like us." In the aftermath of September 11, 2001, we are experiencing a new urgency in the need to work out the proper principles of self-defense. The lessons we draw from domestic self-defense and from international humanitarian law should aid us in this intellectual and moral quest for the right rules in this era of new dangers.

3

A THEORY OF LEGITIMATE DEFENSE

1. SELF-DEFENSE AND DEFENSE OF OTHERS

Although all common law jurisdictions refer to the concept of self-defense, the term, at least in this form, is not universally used. Indeed, most other countries on the Continent replicate the German pattern of speaking about "necessary defense" instead of speaking of self-defense. The French term, which is reproduced in all other Romance languages, is simple and direct: *légitime défense,* or legitimate defense.[1]

The differences between the English concept of self-defense and the French concept of légitime défense are deep and far-reaching. In domestic legal systems, self-defense is supplemented by a whole range of parallel concepts. To use the terms and code sections of the U.S. Model Penal Code, self-defense (§ 3.04) is supplemented by the defense of other persons (§ 3.05) and defense of property (§ 3.06). All three of these forms of defense would be covered by the French concept of légitime défense. The three Model Penal Code sections employ a total of 1,881 words to micromanage the range of issues that are covered by the typical European codes' provisions on legitimate defense in a single sentence.

For example, the German Penal Code Article 32 holds that a "defensive use of force is legitimate when necessary to avert an imminent unlawful attack to oneself or others." The attack on oneself or others includes all interests of the person: bodily security, property, and even intangible interests such as privacy. How is it possible that this single sentence adequately covers the same territory of defensive force as the 1,881 words of the Model Penal Code? The answer is simple: it is not meant to. Criminal codes in the Continental tradition are not recipes for solving all possible problems. Rather, they state general principles and leave the rest to be worked out by the case law and the scholarly literature.

The distinction between self-defense and légitime défense also exists in international law, although here the confusion is even greater. The French-language version of the United Nations Charter, which is just as authoritative as the English version, uses the phrase *droit naturel de légitime défense* to refer to what is termed in the English version "the inherent right of self-defense." The "inherent right" becomes a "natural right" in French; even more intriguingly, "self-defense" becomes the more generalized concept of légitime défense. Parallel translations are found in Spanish and Russian.[2]

It is difficult to know, therefore, which term, self-defense or légitime défense, we should use from this point on. Both terms carry official weight in international law; both are used in official versions of the UN Charter. But Article 51 also uses another term that muddies the waters. Alongside individual self-defense, the Charter mentions "collective self-defense," which adverts to one nation assisting another in warding off an aggressor. For example, in 1991 the United States engaged in collective self-defense by going to the rescue of Kuwait against Iraq. There is, then, a potential for overlapping references to the right of states to rescue each other from aggression. This form of defensive force could be treated as collective self-defense, or it could be implied by the defense of others included in légitime défense and even in some statutory definitions of self-defense.[3] However, as it is described under Article 51, the right of states to come to each other's aid is freighted, in international law, with concerns about covert aggression, or coming to the aid of another and then remaining as the occupier of the defending country.[4]

These fears complicate the question of principle. Should states have the inherent or natural right to defend not only themselves but other states that are subject to aggression?[5] If Kuwait is subject to attack, do all states have a right to send troops to defend it? If Vietnam is subject to attack from the North, may Americans send troops even without the consent of the South Vietnamese government? Americans assumed that they needed to be invited by the sovereign government in Saigon.[6] But this interpretation is not self-evident. If we follow the French charter text and instead assume a right of légitime défense, there is little doubt that saving other states from aggression falls within the natural right recognized by the (French version of the) Charter.

The enormous difference between these two concepts of defensive force—self-defense and legitimate defense—shapes the assumptions of international law. Because most scholars follow the English rather than the French text, they have tended to assume that this "inherent right" is limited to the protection of one's own borders. Insofar as our choice of language determines policy, we should perhaps be more conscious of our choices.

We will not yet take a stand at this point as to whether an inherent right of individual or collective self-defense under Article 51 should include the automatic right to come to the defense of other states subject to aggression; this question of law and policy requires further discussion. It can only tantalize us at this juncture, for it illustrates a major theme of comparative law that lies at the foundation of our inquiry: that we should seek to refine international law by drawing on the lessons of other national legal systems, including their domestic law of self-defense. However, we must first come to grips with our two opposing traditions of legislation: the Continental conception of the Right and the truncated notion that dominates English-speaking common law jurisdictions.

Is language destiny? Does our choice of English or French or German determine whether the inherent, natural right protected under Article 51 of the Charter includes the automatic right to protect other states? No, language is not destiny, but sensitivity to comparative law should make us aware of the way our choice of language limits our horizons. Just because we speak English and this book is written in English, and the English text of Article 51 speaks of "self-defense" instead of légitime défense, we should not assume that the Charter approves only of the force exercised in the name of protecting a country's own borders. For the time being we will bracket this intriguing question of the diversity of language and conceptual frameworks in international law as we focus on a deeper understanding of légitime défense and the use of force among nations.

2. THE UN CHARTER AND LÉGITIME DÉFENSE

What is the proper analogy for thinking about Article 51 and its promise that member states have the right to use military force when they are attacked? One might, for example, analyze this right in the same way one might interpret a right found codified in a regular business contract.[7] Think of a contract that imposes responsibilities for each party but also contains a provision enumerating the rights of each party in the event of a breach by the other. One might read Article 51 in this way. Signatories to the UN Charter have agreed to forgo the use of force by virtue of their signing the "contract," and if a signatory fails to live up to its obligations, member states who are attacked are relieved of their contractual obligations and can engage in self-help measures. Under this view, the proper analogy for reading Article 51 is not criminal law at all, but private law.

This view is enticing but fails to capture the intuitive appeal of viewing Article 51 through the lens of criminal law. The reference to self-defense in Article 51 is a justification for otherwise illegal violence for the simple reason that self-defense has always been a defense in international law for using military force, long before the UN Charter was signed. As will become clear later in this chapter, the major treatise writers of international law long ago recognized the intuitive appeal of self-defense and self-preservation, terms that originated in criminal law but can be used whenever defensive force is applied against aggression, whether individual or collective. So if criminal law is the appropriate lens through which to interpret Article 51, how specifically do we critique the rather concise provision recognizing the inherent right of self-defense?

Although the English-language version of the UN Charter refers to collective self-defense, the equally authoritative French-language version uses the phrase "droit naturel de légitime défense" to refer to what is termed in the English version "the inherent right of self-defense." The first thing one notices about this language is that the right is inherent. Many things could be meant by the phrase "inherent right," although the French language is clearer: the right is natural. This means that the right is not created by the UN Charter but is preexisting and is simply recognized by the Charter. Because it is an inherent right, it should not be limited to member states of the United Nations. In fact, it is unclear whether it must be restricted to states; it might apply to all nations or peoples, in the broadest sense, regardless of whether they are in control of their own state or not. This is an important aspect of légitime défense and is explored in greater depth in chapter 6.

But we must set aside this question and deal with the more central issue of how to understand the term légitime défense. Should states have the inherent or natural right to defend not only themselves but also other states that are subject to aggression? If Kuwait is subject to attack, do all states have a right to send troops to defend it? If we follow the French charter text and assume a right of légitime défense, there is little doubt that saving other states from aggression falls within the natural right recognized by the French version of the Charter.

The scholarly literature on the use of force in international law is generally insensitive to the nuances of the French-language provision of légitime défense. The major treatises on the use of force rarely mention the phrase légitime défense or legitimate defense at all, and when they do, there is no discussion of its unique differences with the Anglo-American notion of self-defense. Consequently, legal interpretations of Article 51 of the UN Charter have remained remarkably static, oscillating between strikingly similar positions that debate minutiae such as when the use of force must be reported to the Security Council and when regional treaties such as NATO can be used to maintain collective peace and security. But the core doctrine of the use of force has remained largely unexamined in the fifty years since it was created in the aftermath of World War II. If the differences between self-defense and légitime défense were properly understood by international lawyers, a serious discussion could begin about the rights and responsibilities of each nation to respond to armed aggression whenever and wherever it occurs.

For example, much of the scholarship has been preoccupied with understanding the notion of collective self-defense. Article 51 posits an inherent right not just to individual self-defense, but also to collective self-defense, a term that has generated much confusion. Though one might think that collective self-defense would encompass our notion of defense of others, a deeper investigation reveals that the concepts are quite distinct. The prevailing view over the past fifty years has been that collective self-defense refers to regional treaties of mutual defense, where several states pledge that an attack against one is an attack against them all, thus resulting in an armed response. These arrangements were once plentiful and even overlapping,[8] although the advent of the cold war led to the rise of NATO and the Warsaw Pact as the only two such arrangements that really mattered. And since the collapse of the Soviet Union, and with it the Warsaw Pact, NATO has grown so large as to encompass half the globe. Consequently, one hears very little about collective security arrangements anymore. Indeed, it was somewhat striking that 9/11 was the first time that NATO declared that one of its members had been attacked, triggering a legal duty of the organization to respond in defense.[9]

Collective self-defense is not the same as defense of others. For example, there is a tendency to reduce all cases of collective self-defense to regular self-defense, assuming that there is always some kind of self-interest involved, such as a regional security interest.[10] Consider the following passage from Dinstein:

> Believing as they do that, in the long run, all of them are anyhow destined to become victims of aggression, each may opt to join the fray as soon as one of the others is subjected to an armed attack. In truth, it is the selfish interest of the State expecting to be next in line for an armed attack that compels it not to be indifferent to what is happening across its borders. It may be said that an armed attack is like an infectious disease of the body politic of the family of nations. Every state has a demonstrable self-interest in the maintenance of international peace, for once the disease starts to spread there is no telling if and where it will stop.[11]

Although this is usually true, it is perhaps not universally true. The argument appeals to the rationale behind these regional security arrangements. They are regional in nature because nations in one geographic area usually face a common military threat. Consequently, an armed attack against a confederate is usually a threat to one's own security, because the attack is evidence of a hostile purpose and may be a prelude to a larger campaign against other countries in the region. Furthermore, the attack against one territory may result in territorial expansion that can be parlayed into greater regional military dominance, in turn resulting in a greater threat to neighboring countries. Therefore, responding to an attack against one regional neighbor is really just an elaborate game of self-defense, because the other members of a security alliance are inclined to protect their regional interests; this is the real reason for their intervention.

This may be true in cases of regional treaty arrangements, but it is somewhat outdated and fails to take into account the modern phenomenon of humanitarian intervention. Many nations feel compelled to intervene (or should feel compelled) when an armed attack occurs, and not just because their regional security interest is implicated. Remember the rhetoric of the first President Bush before Operation Desert Storm. He called the invasion of Kuwait an act of aggression that could not stand. It was not an attack against U.S. interests; it was an attack against a sovereign nation, and Bush argued that a full-blown invasion of a sovereign nation was simply intolerable in this day and age. Of course, the U.S. military action was authorized by a Security Council vote under its Chapter VII powers. However, the Bush administration would most likely have sent troops even if the Security Council had failed to act. And the Bush administration's stated reason was a normative argument that Saddam Hussein's actions were wrong and Kuwait had to be defended. Although commentators and skeptics made much about the U.S. interest for cheap oil—and stability in the Middle East—this was never a part of Bush's early rhetoric about the rationale for the war.

The existing legal literature on collective self-defense fails to deal with the most central issues raised by the concept of defensive action. The literature overemphasizes an odd collection of legal issues: regional treaty arrangements for mutual assistance, concerns about aggression and unlawful occupation, and the duty under Article 51 to report actions of self-defense to the Security Council. International lawyers worry that regional treaty arrangements will be used as a legal strategy to defend military aggression. Small cases of provocation or border incidents will be manufactured to provide the legal justification for a military attack that was already planned well in advance. This concern is probably a holdover from the First World War, when Archduke Ferdinand was assassinated and a set of overlapping alliances quickly spiraled into a world conflict. The more modern version of this phenomenon is the Gulf of Tonkin incident, where the United States used a naval attack in the Gulf of Tonkin as a legal basis for greater military involvement in the region.[12] Evidence collected in the meantime has suggested that this supposed attack was in actuality a deft pretext for war.[13] These cases of questionable defense might also lead to unlawful occupations, where a stated concern for the continued security of the state is used as a pretext for the continued occupation of foreign soil, long after the alleged justification for the attack has passed. These concerns were paramount during the cold war, when the Soviet Union intervened, with military force, to create the Eastern Bloc. Although ostensibly these interventions were motivated by a desire for collective self-defense, the Soviet presence in reality became an unlawful and sustained occupation that continued until the collapse of Communism.

All of these issues sidetrack the needed discussion of humanitarian intervention, which we take up in greater detail in chapter 6. When states engage in a humanitarian intervention, they are hardly acting in collective self-defense. The relevant questions are not regional treaty arrangements and a legalistic duty to report the use of military force to the Security Council. Rather, the pertinent issue is whether the humanitarian crisis is legally sufficient to justify a violation of another state's territorial integrity on the basis of defense of others. This would be central to our discussions if we switched from talking about collective self-defense and started talking about légitime défense.

Analyses about collective self-defense also place too much emphasis on whether the target state *believes* that it is being attacked. The International Court of Justice concluded in the *Nicaragua* case that for a defending country to exercise the right of collective self-defense to aid an attacked country, the latter must conclude and announce that it has been attacked.[14] In other words, the defending state cannot rely on its own judgment about whether an attack has occurred and whether a military response is required. More significantly, the attacked state must formally *request* the assistance, according to the ICJ, or else the exercise of collective self-defense is illegal. This is a startling claim. In many cases, there is no time to wait for a thorough analysis and formal request from the attacked state. Also, the government of the attacked state may be seized by sympathizers or collaborators of the attacking state, who would refuse to issue such a formal request. One can easily imagine this happening in cases of ethnic conflict, where an attack is initiated and at the same time the government of the attacked state is taken over by members of an internal ethnic group aligned with the attacking state. In such a case there would be a real problem of determining whether those who have seized control are a legitimate government or not. If the government is not legitimate, are all other states barred from exercising collective self-defense because a formal invitation is lacking? But the real question to ask is whether the need for a formal invitation in cases of collective self-defense is appropriate, and whether switching to légitime défense would vitiate this requirement. Indeed, the dissenting judges in the *Nicaragua* case noted that the invitation requirement was impractical.[15]

Dinstein's analysis of the *Nicaragua* case assumes that the ICJ went wrong, but for the opposite reason. According to Dinstein, when one state intervenes to defend its neighbor, it does so to protect its own interests and is therefore acting in some form of self-defense. This could be one reason to consider relaxing the formal assistance requirement because we would not want to handcuff the responding state's right to use military force to protect its own interests. Although the conclusion here might be correct, the reasoning is not. In some situations, the reason to relax the requirement of a formal request for assistance is that the responding state is engaging in defense of others, not some contorted version of self-defense. Just as anyone on the street has the right to come to the defense of a victim of a violent attack and need not wait for a formal request, so too any member of the world community should have the right to engage in defense of others when a violent attack has occurred. The continental notion of légitime défense recognizes this fact because it encompasses defense of others.

Consider the case of Austria.[16] The Nazis walked into Austria in 1938, behind the threat of invasion but without any actual resistance. Had the Austrians resisted, they clearly would have been invaded and slaughtered. It is therefore incorrect to call the Anschluss a case of voluntary political union between two countries. While everyone agrees that other countries had the right to intervene against the Nazi territorial drive, there are different rationales that one might appeal to. Dinstein would appeal to self-defense proper, arguing that neighboring states had a right to intervene to protect their own regional interests because they certainly knew that they were next on Hitler's radar. Proponents of collective self-defense are inclined to ask questions about the existence of formal treaties of mutual assistance and whether military intervention is required under them. But both avenues totally miss the point. Every country had the right to intervene on Austria's behalf, whether they asked for help or not, under a theory of defense of others. This is clear under our notion of légitime défense. If the Austrian government was seized by Nazi sympathizers who retroactively approved the Anschluss, would this change matters? It should not, because territorial expansion by military force, or by imminent threat of military force, is intolerable in international law. And all nations have the right to come to the defense of others in such circumstances.

A more contemporary example will help explain the point. After Saddam Hussein brazenly invaded Kuwait in 1990, the first Bush administration declared that Hussein's illegal aggression could not be allowed to stand. What was the basis for the intervention? Is it best described as a case of collective self-defense or a case of defense of others? Although clearly Bush argued that Hussein was dangerous, and clearly the United States had strategic interests all over the Middle East and the Arab world, it is nonetheless clear that it was Kuwait's defense that we were coming to. Hussein had engaged in all sorts of threatening military behavior for decades, including a long war with Iran that the United States supported, but it was not until Hussein started invading sovereign nations that the first Bush administration decided to act. We acted on behalf of Kuwait and defended its right to exist. Although elements of both collective self-defense and defense of others are present in this example, it more closely hews to the core idea of defense of others. And any account of international self-defense that excludes defense of others in favor of collective self-defense misses the point. Intuitively, the concept of defense of others and légitime défense more accurately tracks what actually happened.

3. THE HISTORY OF THE UN CHARTER

The negotiating history of the UN Charter complicates our analysis of the meaning of self-defense and légitime défense. One might think that the delegates deliberately chose the words "individual and collective self-defense" and rejected "defense of others." If they had really meant to enshrine légitime défense as the appropriate standard, they would have chosen a corresponding English phrase that included defense of others. But this argument does not display a subtle understanding of the negotiating process and the genesis of the Charter. First, the initial negotiations focused on creating an English-language document that every nation could agree on. However, the fact remains that the French-language version was adopted at the same time as the English version and carries equal legal weight. It is neither derivative nor secondary. Rather, it appears most likely that insufficient attention was placed on the translation process. Indeed, a historical analysis of the negotiations indicates that the wording of Article 51 was some of the most contentious in the whole process and remained a sticking point until compromises could be brokered. A nuanced reading of this history indicates that the exact contours of the right to self-defense depend on whether you look to the French- or English-language version of the Charter.

It was clear during the negotiations that the UN was to be the clearing-house for all decisions about the use of military force and that the Security Council was the appropriate organ to make these determinations. Indeed, anything less would have threatened the UN's status as a strong international institution capable of responding to threats to international peace and security. However, concentrating so much legal power in one organ—with limited representation of the world's nations—caused much controversy. In particular, Latin American countries were concerned that the Charter could be used as a legal justification for invalidating their regional treaty arrangements for mutual security.[17] Under these agreements, an attack against one country would trigger a duty to assist by the remaining countries. In the wake of World War II, Nazi and Japanese aggression, and the failure of the League of Nations, these treaties for mutual security were obviously important for the world's minor powers, and most were reluctant to sign on to any international legal order that would have invalidated them—at least until they could be certain that the global institution that would replace them, the United Nations, would be successful in adequately protecting their security and preventing another world war.

There was also great concern that the veto power held by the Major Powers on the Security Council would prevent the Security Council from authorizing military action, either when one of the Powers was itself responsible for the aggression or when it was in some other way aligned with a minor nation that committed the aggression. A single holdout among the Great Powers could then handcuff the Security Council from acting. In light of this fact, the regional security treaties were seen as an important legal fallback that could be used to justify military action in the wake of Security Council inaction. If the Security Council did not authorize the use of force to respond to an illegal aggression, at least an organization of like-minded nations with a common regional security interest could step in to fill the void.

The Latin American nations were therefore inclined to support an amendment that carved out an explicit exemption for these regional security arrangements. In other words, use of force by these regional organizations would also constitute a legal use of force under international law and would not be subject to the requirement that use of force be authorized by the Security Council. The United States was understandably concerned that this amendment would create an exception so large as to weaken the international institution at the hands of regional interests.[18] If regional organizations could authorize the use of force, nations would have a greater incentive to sign mutual defense pacts and less incentive to participate meaningfully in an international institution for peace and security. Indeed, this was precisely the kind of alliance that proved disastrous in the lead-up to the First World War and would "convert the world into armed camps."[19]

The French delegation responded to this concern with a very particular suggestion. Instead of exempting regional defense treaties, the UN Charter should explicitly state that all members had the right to act in the interests of "peace, right, and justice" in the event of Security Council inaction.[20] The precedent for this proposal was Article 15 of the Covenant of the League of Nations, which stated that "the Members of the League reserve to themselves the right to take such action as they shall consider necessary for the maintenance of right and justice."[21] In many ways, this was a startling proposal and represented a totally different worldview of the relationship between the UN and its member states. Under the French proposal, each state would be responsible for making individual determinations of when intervention was required by the principles of what was just and right—obviously a very broad standard—and could not be handcuffed in its determinations by a failure to act at the international level. In some sense this proposal merely codified the de facto status quo, because when individual states believe that intervention is necessary and the Security Council fails to act, they intervene anyway. The French proposal would simply have provided a legal basis to justify these actions instead of always labeling them illegal under international law. Each member state would bear some ultimate responsibility for deciding when justice demanded the use of force.

The United States was so wary of providing explicit exemptions for the regional treaty arrangements that initially it supported the spirit of the French proposal.[22] This was a better solution than turning the globe into regional "armed camps." But nations that relied heavily on mutual defense pacts, particularly the Latin American states, were interested in a more explicit endorsement of their security pacts rather than the loose language of the French proposal.

Several compromise amendments were considered that ultimately paved the way for the final UN Charter. First, the language on the right of self-defense was changed so that it would be characterized as "inherent."[23] Second, it was the British delegation that suggested the compromise that would solve the problem of the regional treaties and would eventually find its way into the Charter. The amendment called for recognition that individual states had the "right to self-defense against armed attack, either individual or collective."[24] When this right was recognized, member states had to report their actions immediately to the Security Council. Although certainly not as broad as the French view that member states could intervene on the basis of justice and right, the compromise was nonetheless broader than it would have been had it recognized only individual self-defense. The reference to collective self-defense was meant to legitimize the regional defense treaties and the kind of collective action they might engage in. However, to further mollify the Latin American delegations, the Charter also included an explicit reference in Article 52 to these regional arrangements, which were deemed legitimate as long as they were "consistent with the Purposes and Principles of the United Nations."[25] Furthermore, these arrangements were to be encouraged so as to promote the peaceful resolution of local disputes.

Although these negotiations had taken place in multiple languages, the focus of the work had been to produce an English-language Charter that all could agree on. It is therefore pretty clear that the delegations did not fully turn their attention to the problems that would be created by translations. As it happens, when legal concepts have exact analogues in another language, the process of translation is uncomplicated. But when a particular legal concept is not represented in another language, translation can be as much of a legal negotiation as a linguistic one. For example, after the Meiji Restoration, the Japanese legal system began to have significant interactions with the rest of the world and required translations of European legal codes. The Japanese, however, did not have a word that directly corresponded to the Western notion of rights, a highly individualistic notion foreign to the Japanese legal system. Scholars responded to this lacuna by simply adopting a new word for the concept of rights: *kenri*.[26] The word first appeared in the Chinese translation of Wheaton's famous nineteenth-century treatise on international law, which was published in China in 1864 and Japan in 1865.[27] The term then migrated from the Chinese language to the Japanese. The concept continues to be used in contemporary Japanese legal culture today, although debate remains about its comparative significance.[28]

However, this example is relatively easy to deal with because a new word could simply be created. A blank slate was available on which a new distinction could be created from scratch. It is far more dangerous and confusing when two languages have apparently corresponding legal terms, but their content is subtly different. This appears to be what happened when it came time to translate the draft UN Charter. The Dutch delegation suggested that there be only one official version of the Charter so that there would be one definitive text to appeal to when there were disputes of interpretation.[29] Although this was correct in principle, the issue of which language would be chosen for the authoritative version was obviously insurmountable. The Soviets would never have accepted an English-only Charter, and neither would many other countries. Although there was no mechanism to deal with disputes of interpretation between the different translations, this difficulty could not be avoided.

All agreed that translations were necessary for each of the official languages of the United Nations, corresponding to the Great Powers on the Security Council. A translating committee was created to take charge of the effort, to be overseen by representatives from multiple delegations to ensure the technical accuracy of the translations. However, this process was clearly not scrutinized to the same degree as the original negotiations. When the French-language version of the charter was created, the inherent right to self-defense became "le droit naturel de légitime défense," apparently without a widespread discussion about the distinctions between the Anglo-American notion and the Continental one. In this case, the fact that there was a corresponding yet distinct concept in another language proved confusing. If the French language had no word for self-defense—just as the Japanese had no word for rights—a new word could simply have been invented. But, as it was, a similar concept already existed that could be used off the shelf, even though pertinent differences remained.

The French had good reason to support the translation of self-defense into their more familiar notion of légitime défense. The Continental notion of légitime défense, encompassing as it did the notions of defense of others and defense of property, was much closer to their original proposal allowing military interventions in support of what was just and right. Indeed, one might even say that the French proposal survived, albeit covertly, in the French-language concept of légitime défense. The concept supports a much broader range of permissible interventions, not just limited to individual self-defense or mutual defense arrangements. Any nation has the right to intervene when nations fall victim to illegal aggression.

It is not surprising that these contradictions were not worked out in advance. Ambiguities are often left in place, as opposed to explicitly resolved, because otherwise no agreement could be reached. The uncertainty functions as a kind of delay in agreement that allows the process to move forward. Take, for example, the question of who would be empowered to interpret the UN Charter. Under U.S. law, the Supreme Court has ultimate authority to interpret the Constitution and adjudicate disputes between the two other branches of government. Anticipating that similar disputes would arise over the UN Charter, some members favored an amendment that would grant the International Court of Justice the authority to issue binding interpretations, and others favored giving this responsibility to the Security Council.[30] When no agreement could be reached, the issue was left unresolved. The result was that a known ambiguity was left in place because each party could interpret that ambiguity in the manner it felt appropriate. No matter that these interpretations are mutually exclusive. As it stands, the issue of who has the ultimate authority to interpret the Charter remains unresolved.[31] Similarly, it was rational for parties not to resolve every last issue of self-defense, one of the most contentious subjects during the Charter negotiations. Instead of resolving the differences between self-defense and légitime défense, instead of defining every exact situation in which the use of force could be used, the members left open the outer reaches of the concept so that some minimal agreement could allow the process to move forward.[32] It is for this reason that the ambiguities about self-defense in international law remain unresolved and in need of deep analysis.

This history of the UN Charter has been largely lost. Few legal scholars are aware of the genesis of Article 51's provisions on the inherent right of self-defense, and consequently, the standard analyses of the provision usually make no reference to its historical origins in the French proposal. Furthermore, the standard texts are equally ignorant of the differences between the English version and the French translation. Whether this stems from historical or linguistic ignorance is unknown.

For example, Thomas M. Franck presents a history of the negotiations in his *Recourse to Force* but emphasizes that the New Zealand delegation proposed an amendment that imposed a duty on all members to "collectively resist" all acts of aggression against any member state.[33] This proposal was rejected, although it did garner a surprising amount of support (a majority in fact), as Franck correctly notes.[34] But the amendment was rejected mostly because it created a legal *duty* to intervene, not because it *allowed* foreign intervention outside of one's borders. The issue was therefore not whether defense of others was appropriate, but whether the creation of a duty to intervene was appropriate. As discussed in chapter 2, this legal issue also arises in domestic law. In the United States, there is no general duty to intervene on someone's behalf unless you have a special relationship with the victim that creates this duty (such as being a lifeguard or bodyguard). Most Continental jurisdictions, on the other hand, have imposed by statute a general duty to intervene. At the international level the issue is also controversial. States were reluctant to agree to the New Zealand proposal because they did not want to be compelled against their will to exercise force against another state, which would in effect transform any local or regional war into a world war overnight. Indeed, much of the UN legal order is built around the voluntary use of force. Even when collective force is used for missions authorized by the Security Council, no state is forced to send troops against its will. Military contributions are made on a voluntary basis, and those who wish to abstain— whether for moral, diplomatic, or practical reasons—may do so.

Franck does take note of the French proposal to allow intervention on the basis of what is right and just, but he takes the rejection of this proposal, combined with the rejection of the New Zealand proposal, as evidence that the Charter's provisions on self-defense are severely limited. This is the standard view. A broad view of self-defense, encompassing defense of others, was defeated when the French and New Zealand proposals were rejected. Regional security arrangements, as a form of enlarged individual self-defense, were approved through the addition of the phrase collective self-defense. No mention is made of the Charter's translation and the term légitime défense. There is no analysis of how the essence of the French proposal, and a wider notion of defense of others, has survived in the French-language version of the Charter, as well as the other translations (such as the Spanish and the German) that follow the French pattern.

The consequence of this omission is certain: the standard legal texts take for granted the primacy of self-defense and totally ignore a more plausible analysis of the provision that remains more faithful to the French-language version of the Charter and the other European translations that rely on corresponding concepts. It is this analysis that we hope to capture in this book.

4. THE NOVELTY OF LÉGITIME DÉFENSE

The current literature on the use of force in international law is strangely insensitive to the distinction between self-defense and légitime défense. Although some scholars are clearly aware of the distinction, particularly those writing in the 1940s and 1950s, when the UN Charter negotiations were still in the recent past, none has realized the import of the distinction or how it might influence their analysis of the use of force. Indeed, even French scholars of international law fail to realize the significance of the distinction, although of course they are aware of the vagaries of translation. Some French authors use the term *le droit d'auto-protection*, literally meaning the right of self-protection, as a way of talking about the Anglo-American notion of self-defense without invoking the much broader notion of légitime défense.

For example, Jean Delivanis correctly notes that the phrase légitime défense spans both criminal and international law and that the concept operates differently in each sphere.[35] But he emphasizes that légitime défense has traditionally operated in a much broader fashion in international law than in criminal law, which is precisely the opposite of our analysis. He says this because he emphasizes the lack of police enforcement power in international law. Self-defense operates in criminal law as a right to defend yourself only until the police arrive on the scene. In international law, the police power is much weaker, and there are few instances

when an international police force will come to a nation's rescue. Therefore, it is true that the sphere of self-defense is somewhat larger in international law. Nations have a greater need to resort to self-help measures
than a random person walking down the street. The response time of the
Security Council pales in comparison to the local police department.

But Delivanis writes some very confusing things about légitime défense,
which is evidence of the general lack of scholarly seriousness about these
concepts, even among French scholars. For example, Delivanis concludes
that collective self-defense did not exist as a *right* in international law
prior to the creation of the UN Charter.[36] In other words, according to
Delivanis, it would be wrong to suggest that the right of collective self-
defense already existed in customary international law and was simply
codified and explicitly recognized when the UN Charter was adopted.

His argument stems from a general confusion among these writers
about collective self-defense and defense of others. They are inclined to
view the two as nearly identical, so that defending others is really just
a case of regional or collective self-defense among a group of nations
that have signed a security pact. But this leads to all sorts of conceptual
mistakes. It leads to the inevitable conclusion that collective self-defense
can be exercised only when a state has some kind of formalized treaty or
obligation to a neighboring state.[37] So, if West Germany is attacked by
the Soviet Union during the cold war, the United States can come to the
defense of West Germany, but only because the United States has signed a
treaty of mutual assistance that requires one to come to the aid of another
if attacked by the Soviet Union. Otherwise, the United States has to sit
this one out.

So far, so good. It is probably correct that a regional arrangement is
required for collective security to be legal under Article 51 of the UN Charter. Indeed, as we noted earlier, the whole point of codifying a right of
collective self-defense in the Charter was to protect the existing regional
treaty arrangements that the Latin Americans wanted to preserve. But
where this analysis goes wrong is in concluding that defense of others
is really just the same thing as collective self-defense. Therefore, a formal treaty of collective security, creating a legal obligation to assist, is
required before the use of force will be considered legal for defense of others too. This is totally wrong, both as a conceptual matter and as a legal
matter. Collective self-defense and defense of others are not the same
concept at all. The former applies when nations form collective security
pacts—this much is true. But defense of others is the notion, more common in domestic criminal law, whereby any law-abiding citizen has the

right to intervene to save someone from an unlawful attack before the police arrive. No formal arrangement need exist. Indeed, the whole point of defense of others is that *strangers* who happen upon the scene can do something about it, even if their only connection to the victim is that they both belong to the great community of humanity.

For these scholars of international law, the confusion on this point stems from an outdated view of defense of others in U.S. law. Delivanis argues that the right of defense of others in U.S. domestic law (*la défense d'autrui*) requires a special relationship with the victim.[38] This was once true at common law, and it was once the view in the United States, but the consensus among criminal lawyers has since evolved, and, we argue, it should so evolve in international law as well. What is even more striking is that the "close relationship" requirement in criminal law never existed in French law at all.[39] It was a species of British and U.S. common law that Continental lawyers in Germany and France rightly avoided. But here we have international lawyers, even French ones, appealing to the U.S. and British rule to influence their analysis of international law.

Delivanis cites Section 76 of *The Restatement (First) of Torts*, the standard explanation of U.S. tort law published in 1934, but this rule of tort law was quickly abandoned in the *Second Restatement* because everyone recognized that the old rule was based on an outdated and offensive view of a master's dominion over his wife, family, household, and slaves. According to the *Second Restatement*, the "restriction of the privilege to intervene on behalf of third persons to those who are members of the actor's family or household was founded upon conditions long since past."[40] The law has long since recognized the right and privilege to defend strangers from attack, and so "every man has the right of defending any man by reasonable force against unlawful force."[41] The case law supporting this change goes back to the early twentieth century; courts have long since recognized that one can defend any stranger who himself has a legitimate right to use self-defense against an unlawful attack.

The rules in criminal law have also evolved. It was once true at common law that a defender could use deadly force in defense of others only when the victim was a family member or an employee. D. W. Bowett, whose *Self-Defence in International Law* offers some of the most sophisticated arguments about the historical tradition of legitimate defense, relies heavily on this analysis of criminal law.[42] He notes correctly that the right of defense of others in English Common Law was limited to cases where there was a special relationship between the victim and the defender.[43] Bowett concludes that "the right is limited to those cases where there exists some sort of proximate interest; the right extends to the defence of one's family, one's servants, and to those persons whom one is under a recognized duty to protect."[44]

Although this was once true, our views about defense of others in criminal law have evolved significantly in the common law world. The traditional view that defense of others required a close family relationship represented an uncomfortable interim moment when scholars were confused about whether defense of others was a justification or an excuse. As discussed in chapter 2, a justification means that the defendant's conduct was justified and there was no wrongdoing at all. An excuse means that the defendant's conduct was not justified but there is some reason that we should not punish him. If the defendant has an excuse, the conduct was still wrongful, but the defendant is not individually culpable for his actions. Defense of others was treated as an excuse because the theory was that you could not stand by and watch while your loved ones were being attacked; you had to do something about it and should not be punished for it when you did. In this sense, defense of others was treated like necessity or duress, where the defendant argued that he should escape punishment because for some extraordinary reason he had to do it. Part of this confusion stems from the fact that the distinction between justification and excuse was largely unknown to common law lawyers, who preferred instead to talk simply about "defenses," a broader category encompassing both justification and excuses without differentiating between them. It was the Continental lawyers, particularly the German criminal lawyers, who understood best the distinction between justification and excuse, between wrongful acts and a defendant's culpability.

And so the close family relationship requirement has now been dropped, even in the United States, for cases of defense of others. It is clear in the Model Penal Code and in all penal statutes that are based on it that defense of others is a justification, just like self-defense. Therefore, there is no reason to treat it like an excuse. We no longer think of the defender as someone who should be *excused* because he could not sit by and watch while his wife was being attacked. (This is the kind of analysis we reserve for the case where the husband commits a crime on an innocent third party because someone threatens to torture his wife.)[45] Lethal force is *justified* to protect victims against unlawful attacks, and the victim does not have to be a relative. Now you always have the right to intervene on behalf of others in the event of an unlawful attack; you do not need to be a bodyguard or a family member. You just have to be *there*, and of course your use of deadly force has to be justified. If a woman is being raped in a dark alley, any bystander can use force to repel the attack; you don't have to be the woman's husband.

In general, there has been a historical evolution away from excuses and toward justifications in international law. Many of the major commentators writing in the early twentieth century preferred to analyze the use of force as a case of necessity that could be excused if the life of the nation was threatened. John Westlake, in his pivotal treatise, rarely mentions self-defense.[46] He prefers to talk of the inherent right of *self-preservation*, which is what excuses the use of military force against another state.[47] He even cites *Dudley v. Stephens*, the famous British case of 1884, where two shipwrecked sailors who ate their cabin boy claimed that their actions were excused by necessity. The judges in the case rejected the defense, and Westlake cites the opinion as evidence that one cannot appeal to self-preservation to extinguish the rights of innocent third parties, whether cabin boys or nation-states.[48] The case is the classic example of murder excused by a claim of necessity, and the great scholars of international law of the day thought of it as the paradigm for thinking about resorting to murder among nations.[49] Going to war could be excused on the basis of self-preservation. The famous *Caroline* case of 1848 describes international self-defense in like terms. British forces destroyed an American ship at Niagara that was smuggling arms across the border into Canada. The ensuing diplomatic incident led U.S. officials to argue that self-defense required a case of necessity, "instant, overwhelming, leaving no choice of means, and no moment for deliberation," a standard that the Americans argued was not satisfied in this case.[50] The idea that one could not do otherwise is the language of self-preservation and excuse. But there has been a gradual shift away from excuses in international law. No one talks about excusing nations anymore. We talk about *justifications* for going to war, because the idea is that if a nation is attacked and goes to war to defend itself, then it has done nothing wrong and does not need an excuse. Its actions are justified.

But some confusion about the distinction between justification and excuses in international law remains. Strangely, the major treatise writers on international law, even some of the French ones, have not fully adapted to thinking of defense of others as a justification. They are still used to thinking of excuses, and this causes them to analyze defense of others as an excuse. They assume that the close family relationship requirement of the common law applied in French and German law as well. Delivanis quotes the French Penal Code provision that allows the "legitimate defense of oneself or others," which says nothing about there needing to be a close relationship between the victim and the defender. Indeed, Article 122–5 of the French Penal Code says, "A person is not criminally liable if, confronted with an unjustified attack upon himself or upon another, he performs at that moment an action compelled by the necessity of self-defence or the defence of another person, except where the means of defence used are not proportionate to the seriousness of the offence." There is no mention of special or family relationships; anyone can exercise this right. In the original French, the code refers to this as *légitime défense d'elle-même ou d'autrui*, meaning literally the "legitimate defense of oneself or others." The close family relationship requirement *never* existed in French law.

Bowett's influential argument therefore confuses collective self-defense with defense of others. None of this would be so problematic except that this analysis of criminal law is used as an analogy to better understand self-defense in international law. But because the original reading of criminal law is outdated and represents a conceptual confusion between justification and excuses, the analogy to international law suffers from the same defects, where they flourish into a poor reading of the UN Charter and the use of force in international relations. Defense of others must be regarded as *part* of a general right of légitime défense, as a reading of the French Penal Code makes clear, and it is not limited to cases of special relationships. It is not an international excuse.

This confusion becomes dangerous when it is imported into international law. Collective self-defense is indeed a restrictive notion, but equating it with defense of others only confuses things.[51] Bowett concludes:

> The "collective" right will exist where the state invoking the right can show some interest of its own which is violated by the attack launched upon another state. We must similarly presuppose a "proximate relationship," though not necessarily in terms of geographical contiguity, which makes a direct violation of the rights of one state an indirect, but nevertheless real, violation of the rights of another state which comes to the former's aid by virtue of the latter's right of collective self-defence.[52]

This just collapses collective self-defense and defense of others into individual self-defense.[53]

Because all of this is based on outdated or incorrect views of French and U.S. criminal law, we cannot equate collective self-defense and defense of others. We must return to the fundamental understanding of légitime défense as a unified concept that allows for defensive force to be used against unlawful attacks—against oneself, another, or property. This is the beauty of the French Penal Code provisions on defensive force, and it is a pity that the conceptual elegance of these terms has been lost on the major commentators on international law. So when the French, Spanish, and German versions of the UN Charter refer to légitime défense and its cognates in Spanish and German, we must understand it within the European tradition as the unified concept that it is. We cannot simply twist it around and make it fit into the Anglophone notion of collective self-defense. This betrays the true meaning of both.

One of the few treatise writers to get this right is Ian Brownlie, who correctly interprets defense of others as a broad right indeed. He concludes that there is "a customary right or, more precisely, a power, to aid third states which have become the object of an unlawful use of force. It is immaterial whether this right is called a sanction, collective defence, or collective self-defence, although none of the terms is really adequate."[54] Indeed, the correct term is légitime défense. Brownlie offers an extensive history of the right of self-defense in customary international law before World War II and the UN Charter. He correctly notes that many commentators on customary international law referred to it as a "right of legitimate defense," showing that the term has a long history in international law.[55] "Legitimate defense," he writes, "involved action to prevent or redress violation of legal rights." It would therefore be correct to say that we aim, with this book, to recapture this seasoned vocabulary and return it to current usage in international affairs.

A deep understanding of the role of the criminal law concept of légitime défense in international law is unknown, not only in the United States, but in Europe as well. This is rather strange, because one would think that the French-language version of the Charter would dictate that the broader notion of légitime défense be applied in the international arena by French lawyers. This is not necessarily the case. Perhaps the field of international law is dominated by English-centric attitudes. In any case, the broad application of légitime défense in the international arena is so novel that even French-speaking countries do not make full use of the concept within their own midst. We aim to rectify that oversight with a coherent account of the doctrine of legitimate defense and its application in international law.

However, we must reclaim this terminology not simply by importing all of its conceptual baggage. We must reclaim the term légitime défense by also associating it with all of the best advances in our thinking about criminal law. The relationship between self-defense, defense of others, and defense of property is more properly understood now as justifications all unified by a common structure. The most basic question is whether the use of force is legitimate, as the French would say, or necessary, as the Germans would say. How do we figure out what is legitimate force and what is necessary? In the great tradition of criminal law, we must analyze cases and determine the constraints that apply to this use of force. Figuring out these constraints—where they come from and how they apply—is the task of our next chapter. Our analogy between international and criminal law will again form the basis for our work.

4

THE SIX ELEMENTS OF
LEGITIMATE DEFENSE

Our starting assumption is that the international law on the permissible use of defensive force should run parallel to domestic law. This point is anticipated in our discussion of the Kantian view that states are moral persons with rights and duties, thus suggesting that parallels are properly drawn between individual and national self-defense.

We begin by paying closer attention to the general principles governing the structure of self-defense in domestic legal systems. In any particular legal system, whether we speak about self-defense, necessary defense, or legitimate defense, we are guided in our analysis by six structural elements. There is no statute or authoritative legal source expressing this consensus, but lawyers all over the world would readily acknowledge the same basic structural elements of a valid claim of self-defense or legitimate defense. We have touched on two of these elements already: necessity and proportionality. In addition, we have alluded to the other two: the requirements that the attack be imminent and that the defender respond knowingly or intentionally to the attack. Now we must assay these other issues systematically, with a view to the analogical relationship between domestic self-defense and self-defense in international law.

The six elements fall into two categories, three bearing on the nature of the attack and three on the requirements of permissible defense. First, the attack must be (1) overt, (2) unlawful, and (3) imminent. Second, the defense must be (4) necessary, (5) proportional, and (6) knowing or intentional in response to the attack. Depending on the particular issue, we shall reason sometimes from international to domestic law, sometimes from the domestic to the international.

1. THE ATTACK MUST BE OVERT

Although the UN Charter overstates the issue, there is some wisdom in Article 51's insistence that self-defense be invoked only when "an armed attack occurs." The drafters meant to underscore the importance of an actual—and not merely perceived—breach of international peace and security. Something must happen in the real world, not just in the minds of those who fear an attack. The conventional, domestic law way to express this requirement is that there must be "overt acts" manifesting the intention to go to war.[1] Massing troops on the border, violating air space, closing shipping lanes, preparing missiles for attack—these are acts that are only preliminary to an actual attack but that manifest a hostile purpose.

The precise actions required are not so clear, either in theory or in practice. One standard of an actual, overt attack is prescribed by the prohibition in Article 2(4) of the unlawful use of force. Here the argument begins with a recognition of the value premises of the Charter, namely, "territorial integrity" and "political independence." Apparently, any use of force that puts these values in jeopardy is unlawful. All states are required to abstain from these unlawful uses of force, with the exception of defensive action under Article 51.

Another approach is suggested by Article 39, which authorizes the Security Council to determine "any threat to the peace, breach of the peace, or act of aggression" and empowers the Council to take a range of measures to restore international peace and security. The relationship between this notion of breach and the defensive use of force under Article 51 is not made clear, but there is good reason to think that the exception for individually determined self-defense parallels the general role of the Security Council in deciding when actual or potential breaches of the peace occur.

It should be underscored, however, that all these standards are hotly contested. For political reasons, the General Assembly cannot agree on a definition of aggression.[2] The main point at issue is whether the legitimate use of force is governed by the political criteria of resisting colonial, racist, and otherwise oppressive regimes. If this point is contested in discussions of aggression, the same problem attends the analysis of "unlawful use of force" under Article 2(4) and "breach of the peace" under Article 39. These issues turn out to be very subtle, but they raise serious questions about the foundations of legitimate force, which we take up in the following chapters. The important point at this stage of the argument is that any attack subject to a response of defensive force must actually, physically occur. There must be overt manifestations of aggressive intentions. The question of whether these manifestations need rise to the level of an "actual attack" will engage us later, when we consider the element of imminence.

The requirement of an overt attack in domestic law is not as obvious as it is in international law. The Model Penal Code has set the tone by emphasizing the *belief* that the actor is subject to an unlawful attack.[3] There is no explicit requirement of an actual attack, provided the actor believes that one is about to occur. There is an additional debate about whether the belief must be reasonable, but that debate runs orthogonal to the problem of where an attack must actually occur.[4] The more fundamental divide lies between those who think that self-defense should be evaluated only against real and objective events in the world, and those who believe that these claims should be evaluated against an actor's subjective beliefs about the world. After all, individuals can make grave mistakes in perception, and sometimes what looks like an attack is not one at all.

Here the Rome Statute is ambiguous. It provides that the actor must "act reasonably to defend himself or herself or another person...against an imminent and unlawful use of force." This seems to imply that there must at least be an overt threat to engage in the imminent and unlawful use of force. Take the case, recently reported in the press, of a U.S. Marine who enters a house in Iraq and finds a wounded insurgent lying on the floor.[5] The Marine is afraid that the wounded man is a decoy lying in wait, and so he kills him. He claims that he has heard many stories of insurgents laying this sort of trap for American soldiers. Is he guilty of the war crime of murdering someone who should have been taken prisoner?[6] Not if he acted in self-defense. He might have "reasonably" believed the wounded soldier was a decoy, but in our view that is not sufficient to invoke a claim of self-defense. The better reading of the statute is that the justification of self-defense depends on whether the defense is reasonably directed against an actual threat—not merely a perceived one.[7]

Self-defense in jus in bello, therefore, follows self-defense in jus ad bellum. As you need an actual attack—at least the beginnings of actual aggression—to justify a defensive response under Article 51 of the Charter, so you need an actual attack, or the objective signs of one, to justify a legitimate response under the Rome Statute.

In this context, international law has something to teach the domestic understanding of self-defense. The writers on self-defense have become so engaged by issues of belief and intention on the part of the defender that they have ignored the requirement of a real-world event manifesting a hostile purpose. One reason we should require an overt act is that killing another human being is too grave an act to be justified solely by the belief—even the understandable and reasonable belief—of the defender. There must be an event that the whole world can see to justify the use of force.[8]

2. THE ATTACK MUST BE UNLAWFUL

It would be hard to find a national statute on self-defense that failed to require that the attack be unlawful. The Model Penal Code, the Rome Statute, and the German Penal Code are all clear on this point.[9] The point of this requirement is to exclude self-defense against justified attacks (which are not unlawful), while permitting self-defense against unlawful attacks that might be excused on grounds of insanity, duress, mistake, and other considerations that deny the culpability of the aggressor. All legal systems are in accord about the importance of the distinction between justification and excuse. Justified actions are not unlawful; excused actions are. If justified actions are not unlawful, no one can invoke self-defense against them. This principle follows from the idea that those who are acting in self-defense are acting lawfully, which implies that they (metaphorically, of course) have the law on their side. In any particular dispute between incompatible interests, the law can be on only one side of the matter. The weight of the law comes down on the side of defenders and against aggressors.[10]

Surprisingly, the element of unlawfulness is not expressly mentioned in Article 51 of the UN Charter, yet it is obviously implied. Consider the use of peacekeeping forces mandated by the Security Council under Article 42 to restore "international peace and security." When such troops enter a country, can their intervention be considered an "armed attack," triggering a right of legitimate force under Article 51? Technically, if you take the words at face value, perhaps so. But the point of the entire Charter is to deny this possibility. Troops dispatched by the Security Council do not act unlawfully. The law is on the side of the Security Council. Their actions prevail over the inherent right guaranteed by Article 51.

3.   THE ATTACK MUST BE IMMINENT

Most legal systems explicitly require an imminent attack. The common law was clear on this point, and German law refers to the requirement of a *gegenwärtigen Angriff* (present attack). The Rome Statute uses the word "imminent," translated as well in the parallel official languages.[11] Only the Model Penal Code deviates from this language by placing the entire burden of analysis on the defense as being "immediately necessary...on the present occasion."[12]

The requirement of imminence means that the attack has not yet occurred, but the defender cannot wait. This requirement temporally distinguishes self-defense from the illegal use of force in two ways. Retaliation against a successful aggressor is illegal because it is force used too late; a preemptive strike against a feared aggressor is illegal because it is force used too soon. Legitimate self-defense must be neither too soon nor too late.

It is fairly easy to see why the use of force may not be too late. If self-defense is exercised too late, after the threat has passed, the only way to justify it would be as punishment for the past attack, or, in the language of international relations, a reprisal or punitive response. But analyzing self-defense as punishment for a previous attack presents serious complications. The reasons are the same in both domestic and international law: because all citizens are of equal authority, no one is entitled to inflict on-the-spot punishment on another. Because all states are equal, no state is entitled to punish another. Indeed, Kant went so far as to claim that punitive wars were a conceptual impossibility.[13] The very concept of punishment presupposes the authority to judge and condemn another, and because no state may do so relative to another, the effort to conduct a punitive action will always fail—not as a matter of fact, of course, but as a matter of principle. Even if states and individuals could logically punish one another, they should not. Just punishment presupposes a process of trial and deliberation; making a judgment on the spot and then executing it is hardly a legal process. Punishment is necessarily an institutional remedy, not one that individuals may take under their own initiative. So if force is used too late, after the threat has passed, it falls into the category of unjustified aggression.

Explaining why the use of force may not be too early is more complicated. One would assume that when the use of force comes too soon, it
should be described as preemptive, but we must exercise care in using this
word. In standard cases of self-defense between individuals, lawyers rarely
describe actual self-defense as preemptive action. When individuals face
the threat of imminent attack—an aggressor points a gun and the victim
resists—the use of force is simply called *justifiable* self-defense.

Yet international lawyers routinely describe responses to imminent
threats as preemptive self-defense.[14] Michael Walzer follows this practice
by describing Israel's Six-Day War against Egypt in 1967 as, at best, preemptive action.[15] He does not mean to condemn the Israeli action, however, and under the facts as they appeared to the world, it would have
been difficult to do so. Egypt had closed the Straits of Tiran to Israeli shipping, massed its troops on Israel's border, and secured command control
over the armies of Jordan and Iraq. In the two weeks preceding the Israeli
response on June 5, Nasser had repeatedly made bellicose threats, including threatening the total destruction of Israel. Most observers concur that
Israel's action was justified under the Charter of the United Nations. In a
rare showing of solidarity in the United Nations, the Security Council—
however quick it may be to chastise Israel's policies—tacitly approved of
its defensive action.

4. THE DEFENSE MUST BE NECESSARY

As three elements define the kinds of attacks that may be resisted by defensive force, three factors circumscribe the permissible range of defensive
action. Force is never justified unless it is necessary, but unfortunately,
necessity is one of the most elusive concepts in our philosophical and
legal vocabulary. Kant wrote, "Duty is the necessity of an action out of
respect for the law."[16] He meant that duty-bound actions are dictated by
duty in the same way that the natural world obeys the necessity of physical laws: if the sun shines on ice, then the ice necessarily, unavoidably,
ineluctably melts.

When self-defense is described as necessary, the word "necessary" does
not mean the same thing as our colloquial understanding of physical or
literal necessity. Nonetheless, the criminal codes of the world refer to
the necessity of defensive force, so a different sense of the word must be
meant here. The German Penal Code is explicit; indeed, under German
law, the defense is called "necessary defense," or *Notwehr*. The New York

Penal Code requires that the actor "reasonably believes" the use of force "to be necessary to defend himself or a third person."[17] Interestingly, the Rome Statute says nothing about necessity in defending persons, but as to the defense of property the Statute demands that the use must be "essential," either for human survival or for the success of a military mission. It seems safe to assume that "essential" and "necessary" function as synonyms in these statutes.

What, then, does necessity mean in this context? It is determined by the possibility of using a less costly means of defense: the attack must be repelled by the least costly method available, and there can be no reasonable alternative under the circumstances. In the *Goetz* case, was there an effective response less drastic than firing the gun at the four feared assailants? Would it have been enough merely to show the gun in its holster? Or to draw and point the weapon without firing? Goetz had twice scared off muggers on the street merely by drawing the gun. This suggests that perhaps shooting to kill was not the least costly choice among the reasonable alternatives in the situation.

There are, of course, many arguments in Goetz's favor.[18] The uneven pace of the accelerating train made Goetz's footing uncertain. During his initial exchange with Canty, he rose to his feet and was standing in close quarters with his feared assailants. Showing the gun in the holster or drawing it would have risked the chance that one of the four young men would take the gun away and shoot him. Gauging necessity under the circumstances turns, in the end, on an elusive prediction of what would have happened if Goetz had tried this or that maneuver, short of shooting. In the end, the telling consideration is Justice Oliver Wendell Holmes's wise aphorism, "Detached reflection cannot be demanded in the presence of an uplifted knife."[19]

A clearer case of unnecessary force is the March 1991 assault by twenty-odd Los Angeles police officers against Rodney King after they pulled him over and surrounded him in a parking lot in the San Fernando Valley. He had no apparent weapon. The police could easily have waited until he fell asleep, or they could have used a sleep-inducing gas to disable him. Instead, they went in with batons swinging. They claimed at trial that he had engaged in an aggressive motion known as the "Folsom roll," so-called apparently because inmates at the Folsom Prison used it against their guards. The country was shocked by the videotaped brutality shown around the world. The early and excessive intervention seemed to be motivated more by a sense of wronged honor than the necessity of using force. Though the four indicted officers prevailed with their theory of self-defense in the state trial, the better interpretation of the officers' motivation was that King had disrespected them by disobeying their orders. The officers were subsequently convicted in a federal civil trial.[20]

Very few scholars pay careful attention to the requirement of necessity, even though it is a universal requirement of self-defense. In the past few years there has been considerable discussion about the duty to retreat when confronted with aggression on the street or at the workplace.[21] The Model Penal Code introduced the duty to retreat,[22] and in the wave of law reform that swept state legislatures in the 1960s and 1970s, many states added this duty to their penal codes. There is no duty to retreat at home,[23] and some advocates of individual rights have argued that the same principle should apply on the street and at work. They are probably right, but changing the law on retreat does not substantially alter the requirement of necessity. Suppose that a handicapped aggressor, confined to a wheelchair, is approaching with a sword pointed right at you. Whether or not there is a duty to retreat in this case, the use of deadly force would not be justifiable. Under the circumstances, it is not necessary to meet force with force. Stepping aside is sufficient to avoid the danger.[24]

In an international context, it would be difficult to argue self-defense if diplomatic means were available to solve the conflict. If Israeli delegates could have sat down with their Egyptian counterparts to negotiate a satisfactory end to the 1967 conflict, the resort to arms would not have been justifiable. In this context, Article 51 of the UN Charter includes an important qualification on the right of states to resort to defensive force; the "inherent" right of self-defense applies only "until the Security Council has taken measures necessary to maintain international peace and security." A corresponding analogy in the domestic sphere might be a cowboy engaged in a shootout with bandits. When the sheriff's men arrive on the scene, the cowboy would be expected to holster his gun and let the state's officials take charge of the situation.

The literature of international law speaks of the requirement of necessity as well as proportionality, but no real sustained attention is paid to either of these criteria. There are good reasons for this indifference to these concepts. Necessity makes sense only relative to the prospect of avoiding and fending off an attack in progress, and the escape of the offender is typically included as an extension of the crime; preventing the offender's escape is considered equivalent to preventing the commission of the offense, as in a famous 1920 German case in which an orchard owner shot and killed an escaping apple thief.[25] The shooting was ruled justifiable as necessary to prevent the successful commission of the crime—but of course there is an issue of proportionality here. For purposes of analysis, however, the two issues must be separated, and in fact the apple thief case is noteworthy precisely because it rejected proportionality as a limitation on the use of force.

It is important to note that in this context the criterion of necessity does not function in the same way as in international conflicts. In the nineteenth century, however, perhaps international lawyers thought it might be possible to prevent attacks from occurring. In Daniel Webster's famous letter to Lord Alexander Ashburton about the sinking of the U.S. ship *Caroline*, Webster claims that defensive force should be justified by "a necessity of self-defense, instant, overwhelming, leaving no choice of means, no moment for deliberation." The British, he claimed, intervened prematurely against what they took to be American intervention on behalf of rebels in Quebec. Today, there is little chance of preventing armed attacks from occurring—except, of course, by striking while the threat is still imminent and not yet operative. If bombers are ordered to strike, if missiles are launched, there is little chance of warding off all these incoming agents of destruction. The most that a defending state can do is to hunker down, take the blow, and then strike in anticipation of the next attack. This is the typical response in international conflict: the defending state uses military force to destroy the military infrastructure of the aggressor nation and thus ward off successive attacks.

The rationale for legitimate defensive force was widely accepted for the U.S. invasion of Afghanistan after 9/11. An attack aimed at the Taliban made sense on defensive grounds, but surely we were not expecting another immediate attack as devastating as the destruction of the World Trade Center. However, this is not to say that there was no threat on the horizon at all. Rather, defensive force was designed to confront the possibility of future attacks that our enemies were plotting against us. There were, to be sure, a variety of future attacks that were in the planning stages, with organizers looking for funding, weapons, and trainees. But this should be carefully distinguished from a specific imminent attack on the horizon. The claim of defensive legitimacy was directed toward possible follow-up attacks in the future, but there were no signs of another attack in the coming days. The exact significance of this distinction between, on the one hand, current or imminent attacks and, on the other, future or planned attacks is discussed in chapter 7.

It is important to recognize that, in contrast to defensive action against aggressors in the domestic context, a military response to attack does not prevent the primary attack, but at best might forestall a follow-up use of force. The intent to avert secondary strikes blends readily with a motive of deterring subsequent aggression by the offending state. The first means of prevention is purely physical; the second seeks to change the motivation of the attacker. The United States and the United Kingdom have recently advanced the argument that their use of force in Iraq was justified not as prevention in the physical sense, but as deterrence,[26] thus starting down the slippery slope of preventive self-defense. The argument of physically neutralizing a not-so-likely second act of aggression is dubious, and the claim of a right to change motivation and deter is worse. If the argument of deterrence were valid, then military attacks could nearly always be justified; it would be too hard to tell the difference between illegal retaliatory attacks and lawful cases of deterrence.

Retaliation is in fact a frequent motive in conflicts, in both international and domestic law. Sometimes the impulse of retaliation is dressed up as a gesture of preserving honor. This was a widely shared sentiment after 9/11: a great power could not simply take a blow like that and not respond. But those who defend the use of violence rarely appeal directly to honor or admit that their purpose is retaliation for a past wrong. The more typical argument is that the individual or nation faces a recurrence of past violence, shifting the focus from past to future violence—from retaliation to defense against an imminent attack. In the case of Afghanistan, the motive of defending the honor of the United States by retaliating was probably uppermost in the minds of the nation's leaders, yet no one dared admit it. The official rationale was, of course, self-defense against future terrorist attacks.

The criterion of necessity makes sense when we are talking about preventing a current attack or frustrating the commission of crimes, but not when the purpose is preventing or deterring successive attacks. What makes the use of force necessary in these circumstances? If bombing Afghanistan was necessary to prevent or deter future attacks—if that is the only factor that justified the bombing—then the argument surely applies against other threats that should be prevented or deterred as well. Under this theory, the analogous use of force against North Korea or Iran would also be justified, in addition to using force against any other nation that might attack us in the future, because we face the possibility of future attacks from them as well. Our actions would be "necessary." Yet Afghanistan was viewed differently, largely because the elements of honor and retaliation made the argument of prevention easier to accept. But if the concept of prevention is doing the justificatory work in the argument, then it should not matter that we were already attacked once by Al Qaeda.

The only thing that would matter is the possibility of stopping future attacks, and those future threats come not just from Afghanistan and Al Qaeda but from a host of different regimes as well. At this point it would be difficult to determine who we can and cannot attack.

5. THE DEFENSE MUST BE PROPORTIONATE

To understand the distinction between proportionality and necessity, let us return to the case of the orchard owner shooting the escaping apple thief. If we consider the prevention of the escape to be an aspect of preventing the crime, then of course it might be *necessary* to shoot. But shooting to kill, or even to wound, would not be *proportional* to the harm threatened. The distinction between necessity and proportionality means that in some cases the use of force might be necessary (no other way to stop the thief) but not proportional (the value of the apple is too small, relative to the taking of life). The disproportion between life and limb, on the one hand, and apples on the other is too obvious to require discussion. No legal regime that imposes a rule of proportionality would treat the shooting of the apple thief as justifiable.

Proportionality in self-defense requires a balancing of competing interests: the interests of the defender and those of the aggressor. Women may kill to prevent a rape. As the innocent party, the woman's interests weigh more heavily than those of the aggressor. But although she may kill to ward off a threat to her sexual autonomy, the situation is more complicated when we ask whether she has a license to take life in order to avoid being kissed or touched. Of course, it matters what kind of touching we are talking about. If the only way she can avoid some truly innocent—but nonetheless unwanted—contact is to kill, that response seems clearly excessive relative to the interests at stake. Even if our thumb is on the scale in favor of the defender, there comes a point at which the aggressor's basic humanity will outweigh the interests of an innocent victim, thumb and all. There is obviously no way to determine the breaking point, even theoretically. At a certain point our sensibilities are triggered, our compassion for the human being behind the mask of the evil aggressor is engaged, and we have to say, "Stop, that's enough."

The rule of proportionality makes sense, but not all legal regimes require it. The German provision does not explicitly require this balancing of the competing interests and has resisted the limitation of proportionality for decades. The traditional view was based on the Kantian theory that the defending party has the right to vindicate his autonomy, whatever the cost. Sometimes this is expressed as a categorical confrontation of right and wrong; in the German expression, *Das Recht braucht dem Unrecht nicht zu weichen* (The party in the right should never yield an inch to the party in the wrong).[27] German writers in the Kantian tradition have reasoned that, as an "ethical matter," the party in the right might stay the hand of defensive force, but there is no legal obligation to do so.[28] In the post–World War II period, German courts have imposed this ethical duty as a legal requirement by importing a principle of private law called "abuse of rights" to help limit a right of defensive force that, according to the Penal Code, extended to all cases in which the use of force was necessary.[29]

The European Convention on Human Rights explicitly limits the scope of self-defense to the defense of persons rather than property.[30] Many American penal codes concur that the attack must represent a threat to the person as well as to interests in property. The New York Penal Law implicitly adopts these restrictions in the principle of proportionality built into its two levels of defensive response: the use of "physical force" and the use of "deadly physical force." The former is permissible only to prevent the "imminent use of physical force" against oneself or against a third person; the more serious response, the use of deadly force, is permissible in specified cases where the threatened force is more serious. Of the scenarios enumerated in the New York Penal Law provision on self-defense, the ones that are most relevant to Goetz's conduct are the threat to use deadly physical force and the attempt to commit a robbery.[31]

The generally recognized right to use deadly force suffers from an ambiguous rationale. Breaking into a house need not, in and of itself, create a risk of harm to the homeowner or to anyone present on the premises. But in some cases such an attack on private premises might entail these risks, and therefore some legal regimes take the view that deadly force is permissible against burglary.[32] Others reach the same conclusion by assuming that property in itself is worthy of the same degree of defense as is applicable in cases of threats to human life.[33] In the international arena, the parallel issue is the use of deadly military force to defend unoccupied territory. Suppose that there were no inhabitants in the Falklands when Argentina attacked. Would the British have been able to defend their interests—even at the cost of Argentine lives? This important issue of principle has received surprisingly little attention from international lawyers.[34]

In applying the principle of proportionality in the international context, we have to distinguish rigorously between two types of cases.

In the one type, the one most like domestic self-defense, the use of defensive force actually prevents the attack. This is atypical in modern warfare, though one could imagine a line of troops blocking the advance of another line of enemy foot soldiers or cavalry. In these cases any necessary means would be permissible, just as in the Kantian model of self-defense. In the second type of case the attack has already struck its target, entailing casualties, and the issue is preventing further attacks or at least containing the continuation of the same attack. In this situation the problem of proportionality becomes critical. For instance, if a single Egyptian bomber reaches Tel Aviv and inflicts a hundred civilian deaths, Israel would undoubtedly respond, but nuking Cairo would be a disproportionate response. The issue of proportionality becomes acute in these cases because the normal requirement of necessity, as we have seen, does little work. Necessity makes sense in the first type of case, where there is some prospect of averting the attack altogether. Where the use of force is directed at presumed subsequent attacks, however, the issues of necessity and proportionality are both in play.

Though there are many different senses of the word "proportional," international law pays almost no attention to this set of distinctions,[35] but they are second nature to good domestic criminal lawyers. In fact, in the analysis of various claims of justification, the concept of proportionality recurs at almost every turn. There are four distinct sorts of proportionality. In cases of self-defense, the disproportion would take the form of *clearly excessive force*, such as shooting an apple thief running off with a single apple. Let us represent this limited exception to self-defense by making it point 90 on a scale of 1 to 100 (Figure 4.1).

To understand Figure 4.1 and the following three illustrations, read the darkened portion as the protection that we afford to aggressors (e.g., apple thieves have a right to life). The blank portion of the graph represents the harm that can be done to an aggressor. Or, expressed from

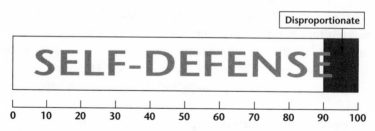

FIGURE 4.1. Disproportionate self-defense

the defender's point of view, the blank portion represents the liberties that the law affords to defenders, for example, the right to respond with deadly force in all but the most extreme cases of clearly excessive force. It would be possible for Figure 4.1 to be all blank. That would be the case if proportionality were not recognized at all as a limitation on necessary self-defense, for example, if we were allowed to kill a thief for stealing an apple. The larger the darkened portion of the graph, the more significantly the principle of proportionality limits the use of force.

The concept of disproportionate force also applies to the traditional justification of necessity, but the meaning of "disproportionate" is entirely different. For example, in the language of the Model Penal Code's provision on necessity (choice of evils), the person using force may act only "if the harm or evil sought to be avoided by such conduct is greater than that sought to be prevented by the law defining the offense charged." Thus, if the disproportion in self-defense is represented as 90 on a scale of 100, the disproportion in cases of necessity would be signaled by the ratio of something like 45 to 55 (Figure 4.2). We get this ratio because the justification of necessity is allowed in cases where the evil sought to be avoided is *greater* than the harm inflicted by the use of force. However, notice that this standard does not require that the evil sought to be avoided be *far greater*. The evil sought to be avoided needs to outweigh the harm inflicted by the use of force only by a little bit before the defense is allowed. As one can see, this kind of disproportionality is far different from the disproportionality we considered above.

These two standards are often confused when assessing, say, the liability of commanders who send in bombers to attack military targets and cause disproportionate collateral damage to nearby civilians. Should the standard for quantifying disproportionality be the limited exception of self-defense, or the larger swath of illegality in necessity cases? This seemingly conceptual point makes a huge difference in the outcome of the

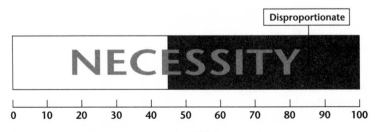

FIGURE 4.2.   Disproportionate justifiable necessity

case, because the ratios on the graphs are so different (i.e., 90 as opposed to 45). Do we assess this case by the self-defense standard of "clearly excessive force," a permissive standard that benefits the defender, or by the standard of "greater harm" that we find in the law of necessity, a standard that allows less room for the defender to operate? Although it is clear that cases of collateral damage in war must be analyzed in terms of disproportionality, at first glance it is unclear if we should apply the version of disproportionality one finds in the law of self-defense or the one found in the law of necessity.

This is a subtle problem complicated by the doctrine of double effect, which holds that if the primary purpose of the action is legitimate (say, the pursuit of a military target), then collateral damage is acceptable provided that the damage is not *disproportionate* to the aim pursued. The underlying moral principle is that the action should be regarded as privileged because of the legitimacy of the dominant purpose. This helps answer the question of what kind of disproportionality is at stake in cases of collateral damage. It is not the sort that applies in cases of necessity as a justification, where the ends and the means are of equal moral value. Rather, we need to apply the notion of disproportionality that places the required harm near the 90 percent mark on our spectrum. It is for this reason that the Rome Statute requires "clearly excessive" harm in its war crimes provision on attacks that damage civilian targets.[36]

As if this situation was not sufficiently complicated, German private law also distinguishes between two different kinds of necessity in cases dealing with property interests. *Defensive necessity* arises when the defender must face a threat bearing down on him by a nonhuman actor.[37] The classic example is shooting a rabid dog. The principles of self-defense apply to human actors who engage in aggression, but not to dogs, who are considered property and are not subject to the norms of the legal system. The interesting feature of defensive necessity (as applied in the case of shooting the attacking dog) is that, as in cases of self-defense, the defender is not required to pay compensation to the owner of the dog (or the victim, in cases of self-defense).[38] Defensive necessity is also sometimes called "passive necessity," to emphasize that the threatening property (in this case, the rabid dog) came to the defender, as opposed to a case where the defender actively seeks out the property that he ends up destroying.

Conversely, there are cases of *aggressive necessity* where the aggressor intrudes upon the property of another to save a higher interest.[39] As typified by a famous American tort case, a ship owner moors his ship to a

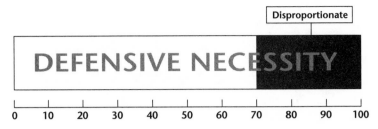

FIGURE 4.3.   Disproportionate defensive necessity

dock owned by another to avoid the devastating impact of an impending storm. The storm crashes the ship into the dock, causing minor damage. In this case, under both German and U.S. law, the ship owner is justified in mooring his ship, but he must still compensate the affected party for the loss of the dock. The reason for the difference is that he voluntarily sought out the dock to protect his ship.[40] Consequently, the standard of necessity differs in this case from the standard applied in cases of defensive necessity. In both cases there is a limit imposed by the principle of proportionality, though each uses a different version of the principle. In the case of defensive necessity (e.g., shooting the attacking dog), the standard lies somewhere between necessity (as a justification in criminal law) and self-defense. It should therefore be represented as point 70 on the graph (Figure 4.3).

In the case of aggressive necessity, though, the standard of proportionality is far more demanding, precisely because the aggressor actively seeks out the property interest that is breached. This standard cuts off the permissibility of the action considerably earlier than the 50/50 line. Let us refer to this as point 30 on the graph (Figure 4.4). In other words, to actively intrude upon another person's property and to cause damage, you have to have a very good reason. In cost/benefit terms, the standard for defensive necessity (shooting the attacking dog) states that the costs cannot be too great relative to the benefits. But the factor of disproportionality is higher than in cases of self-defense because the attacker is not a human being acting in violation of legal norms. In cases of aggressive necessity, where the actor intrudes upon the property of another and causes harm, the benefit must actually be much greater than the costs. It must be a matter of saving life relative to the cost of doing damage to property interests. (This makes sense because an intrusion against someone else's property rights is always questionable, and therefore subject to tighter restraints.)

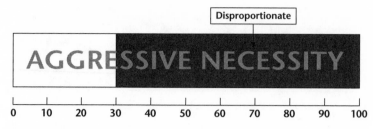

FIGURE 4.4.   Disproportionate aggressive necessity

If these last two graphs are puzzling, think of it in this way: In a case of aggressive necessity, the threshold of disproportionality is set at a lower level, thus the ratio of 30:70 instead of 70:30, as in the case of passive necessity.

These differences reflect two value commitments. First, we should maintain a high respect for private property, even though we have less respect for property if it is a source of risk to others (i.e., the rabid dog). However, if the aggressor selects the victim and deliberately invades the property of another (as in choosing to dock without permission in order to protect the ship), the invasion against the rights of another is more palpable. Second, passive victims of threats have greater rights than those whose body becomes the instrument of harm. If a fetus growing inside a mother's womb threatens her life, the principle of defensive necessity would apply to justify abortion of the fetus—provided, of course, that the death of the fetus is regarded as being within the realm of proportionality. In this scenario, it is of critical relevance that the fetus's moving in the mother's domain is described as a threat. If the fetus were in its own domain, apart from the mother, it would be much more difficult to justify killing the fetus to generate a benefit, even a life-saving one, for another person.[41]

There is another way of expressing these distinctions. If a thing is "tainted" because it generates a risk to others, we generally grant it less value in the balancing process, and this lesser value is reflected by the fact that it becomes possible to apply defensive necessity (as opposed to necessity as the lesser of two evils or aggressive necessity).[42] That being the case, we are left with a serious problem: Why, as a matter of principle, should it matter whether or not an object is a source of risk in a modern rational system of criminal law? Literature in this area has not yet generated a convincing answer.[43] Still, though the distinction between aggressive and passive necessity is not recognized in either Anglo-American criminal law or international criminal law, it remains a valid option that appeals to some scholars.[44]

6. THE DEFENSE MUST BE AN INTENTIONAL
OR KNOWING RESPONSE

The preceding two characteristics of the defense—necessity and proportionality—speak to the objective characteristics of using force under the
circumstances. To assay these issues, we need not ask any questions about
what the defending individual or state knew about the attack and the
defense. In the typical case, the defender is fully aware of the circumstances of the attack and chooses to respond with defensive force. But, to
take a variation on the *Goetz* case, assume a white passenger is surrounded
by four black youths on the subway. No one has made a threatening move,
but in fact the youths have surreptitiously drawn their knives and are
poised to strike against the "other" in their midst. All of this is unknown
to the passenger, but at that moment—for his own reasons—he pulls his
gun and fires with the sole aim of inflicting harm on the four black youths
surrounding him. Only after they have been disabled is it established that
they had knives in their hands and were presumably ready to attack. In
this variation of the case, could Goetz invoke self-defense on the ground
that his act did, in fact, frustrate the attack? It would be a de facto act of
self-defense, even though Goetz had his own reasons for shooting.

The consensus among Western legal systems is that to legitimately
invoke self-defense, the defender must know about the attack and act
with the intention of repelling it. Why should Goetz receive the benefit
of a justification if he acted maliciously, without knowledge (or fear) of
an attack?

Surprisingly, some leading scholars think that in a case of criminal
homicide, the accused should be able to invoke self-defense even if he
does not know about the attack.[45] The scenario would look like this: criminal A plans a murder and starts to commit the crime. Unbeknown to
him, though, his victim B was about to attack him first with a weapon.
Had criminal A known about this attack, he surely would have been justified in using defensive force to stop it. The only problem is, criminal
A knew nothing about the attack and instead killed his victim in cold
blood. Knowledge of the coming attack played no role in his reasoning
at all. However, some legal scholars argue that such a criminal should be
able to plead self-defense anyway. They argue that, if you cannot be guilty
of homicide for killing someone who is already dead (no matter what
your intent), then you should not be guilty of homicide for killing an
aggressor (no matter what your intent). No harm, no crime—and there is
arguably no harm in killing an aggressor. To put the best possible face on

this position, we might think of the injured aggressor's right to complain. Suppose the police arrest someone without probable cause but, though the police acted illegally, the suspect is in fact guilty. Should he be able to complain about the police arrest when they fortuitously got it right? In the context of the Iraq invasion, suppose that the United States gets lucky and finds weapons hidden deep in a cave in Iraq. If we assume that the United States did not have a reasonable expectation of finding weapons in the first place, should its subsequent good luck justify its conduct?[46]

There is a tendency to argue that the guilty party (e.g., Iraq, if it had WMD) has no right to complain—in Iraq's case, to the international community. This might be true, but the community as a whole may also condemn any use of force that is not motivated by the proper reasons. It seems counterintuitive to call someone an aggressor "in mind," and then to allow him to benefit from discoveries made after the intended aggression is complete.

Scholars of criminal law who favor giving the aggressor the benefit of this good luck argue that he should nonetheless be guilty of an attempted killing. In such a case, the killer was acting for malicious reasons, though it turns out, after subsequent investigation, that he would have been justified by self-defense. On this view, the killer is neither fully culpable, like a murderer, because of the existence of objective circumstances that would have justified self-defense. But on the other hand, the killer is not fully absolved, like a bona fide defender, because the objective circumstances were completely unknown to him and he acted for malicious reasons. The killer's culpability might lie somewhere in between, which is why some scholars are tempted to shove this case into an intermediate category like attempted homicide. He attempted to commit homicide, but in some sense he was frustrated from meeting the legal requirements of a murder. There is some support for this view in German law, and Paul Robinson has defended it with some success in the American literature.[47]

This theory could be applied in the international sphere. That is, in the hypothetical case where the United States fortuitously discovers WMD after attacking Iraq, it would then be guilty only of "attempted aggression" against Iraq. Although the phrase sounds incongruous, the theory should be evaluated. One might wonder how the United States could be guilty of attempted invasion if, in fact, it had already landed over a hundred thousand troops on Iraqi soil and caused widespread death and destruction. Attempts are cases of trying and failing to realize the criminal plan. How can it be said that, in our variation of the *Goetz* case, the shooter tried but failed to hit the youths, when they are in fact lying bleeding on the ground? If weapons are found in Iraq, can we really say that the United States (which did not have an *ex ante* justification for the invasion) is guilty only of an attempted invasion of 250,000 troops in Iraq?

The argument for Robinson's position, amended for international law, is very subtle. In his scheme, the crime would not be invasion per se, but *unlawful* invasion. The aggression is supposedly not unlawful if weapons of mass destruction are found. The United States might have the intent to commit an unlawful invasion—that is, to invade regardless of whether weapons are found—but that becomes impossible under the actual circumstances of the case. If the weapons are there, it supposedly would become impossible to commit an unlawful invasion. Therefore, the United States is guilty of attempting to commit an unlawful act. This thesis testifies to the ingenuity of the legal mind, but it runs afoul of our understanding of terms like "trying" and "attempt."[48]

A better rule might be that self-defense is a privilege that can be properly exercised only by people who know the relevant facts. The Model Penal Code stresses the element of belief as a condition for any claim of justification. The New York Penal Code explicitly requires that the actor "reasonably believes" that the circumstances warrant defensive action. The implication is that the defendant cannot take advantage of circumstances beyond his knowledge and intention. The ramifications for the *Goetz* case are far-reaching. Ramseur and Cabey were carrying screwdrivers in their coat pockets; some media even reported that the screwdrivers were sharpened. The reports confirmed what many people wanted to believe: that these four kids were about to mug Goetz and therefore his response was justified as self-defense. Yet there was no evidence that Goetz knew about the screwdrivers. Not one of the four victims pulled a screwdriver from his pocket, either before or during the shooting. And if Goetz was not aware of the screwdrivers, then they had no bearing on his claim of justification. They had no more legal relevance than would a secret plan to kill Goetz because he was white.

The same rule should apply in the international context. The United States cannot claim justification for invading Iraq unless it is motivated by facts "reasonably believed." The claims about WMD must have been believed, and as we shall see later, these beliefs must have been based on public evidence accessible to all. The relevant question, therefore, is not justification *ex post*, or after the action, but justification *ex ante*, on the basis of facts held to be true before the invasion. The real question in Iraq is whether the Bush administration had enough credible, publicly revealed evidence to believe they would find something that had eluded the United Nations inspectors. President Bush might have believed that he would find something, but, as in the *Goetz* case, the real problem was whether his belief was reasonable on the basis of the evidence publicly available. (Precisely why the belief must be reasonable and based on public evidence are matters that we take up in greater detail later.)

The requirements of imminence, necessity, and intentionality are found in numerous legal systems, as well as in the international law of armed conflict; proportionality, however, is a more contested requirement and deserves greater development. In any event, what we have offered in this chapter is not the final word on these concepts. Rather, we have simply introduced their broad outlines and explained their importance for any good theory of self-defense. The full richness of these concepts will be brought forward in full color in the following chapters, as we apply our theory of legitimate defense to the most pressing problems of international law: aggression, humanitarian intervention, and preemptive war.

5

EXCUSING INTERNATIONAL AGGRESSION

1. THE PSYCHOTIC AGGRESSOR

As part of our systematic comparison of self-defense in domestic and international law, we need to address another distinguishing feature of Western legal systems: the availability of defensive force against excused aggressors. This concept may not be obvious to international lawyers. The specific example discussed in the literature of legal philosophy and criminal law is the use of defensive force against a "psychotic" aggressor. Here is the problem from an article written by one of us thirty years ago:

> Imagine your companion in an elevator goes berserk and attacks
> you with a knife. There is no escape: the only way to avoid
> serious bodily harm or even death is to kill him. The assailant acts
> purposively in the sense that he rationally relies on means that
> further his aggressive end. He does not act in a frenzy or in a fit,
> yet it is clear his conduct is nonresponsible. If he were brought to
> trial for his attack, he would have a valid defense of insanity.[1]

In more general form, the problem is whether self-defense applies against excused but unjustified aggression. If the aggressive conduct is itself justified, say, in the case of a police officer effecting a valid arrest by appropriate means, it is clear there is no right to use defensive force in response. If the UN Security Council uses force in accordance with Chapter VII of the Charter, the nation invaded may not resist. Yet if the aggression is merely excused by necessity, duress, insanity, or mistake, the aggressor is not acting as a matter of right or privilege. There may indeed be a right to defensive force in response. The clearest case is force that is excused on grounds of insanity. No one thinks that the insane aggressor is justified, and therefore it is plausible to describe the insane aggressor as acting wrongfully and unlawfully.

There are two basic quandaries posed by the case of the psychotic aggressor. First, if the victim of the attack defends himself and kills the aggressor, should he be acquitted? Second, if a third party, a stranger, intervenes on behalf of the victim and kills the aggressor, should he be acquitted? The first question is relatively easy. It is hard to see either the justice or the efficacy of punishing someone who kills for the sake of self-preservation. The more difficult issue is whether third persons should be allowed to intervene without risking criminal conviction. If one party to the affray must die, either the insane aggressor or his victim, why should an outsider be encouraged to choose sides? Neither is morally at fault; neither deserves to die. Yet it is hard to deny the pull in the direction of favoring the victim of the attack and permitting intervention to restrain and disable the aggressor.

These issues have great ramifications for necessary defense in international law. Probing the puzzle of the psychotic aggressor raises a host of new questions that are rarely addressed in the literature of international law. Can aggression by states ever be excused? What do we mean by excuses as opposed to claims of justification? Is there a difference between applying these concepts to individual behavior and adapting them to the behavior of states? These are difficult questions for both domestic criminal lawyers and international jurists.

Since the original article just cited appeared in 1973, the psychotic aggressor has become a fashionable item of discussion in philosophical circles, largely because Robert Nozick cited it in a book published in the mid-1970s.[2] Kent Greenawalt responded to many of the arguments about justification and excuse in an important article of the mid-1980s, and one of us followed up with further reflections about the psychotic aggressor in 1992.[3] We draw on these articles now in reviewing the controversy and extending the argument to international armed conflicts. This hypothetical case has vast repercussions for the law of defensive force against international aggression.

The problem has been to devise a theory that would account for our intuition that we may resist an excused but unjustified aggressor, typified by the purposive psychotic attacker in the elevator. The search for a solution has focused on the areas of self-defense and necessity, so we turn now to a comparative study of these defenses as they apply to the problem of the psychotic aggressor. As we shall see, there are not merely two rationales for acquittal expressed in the terms "necessity" and "self-defense"—there are no fewer than five, and one of our tasks will be to sort out these different rationales and show how each manifests itself both in common law and Continental legal theory.

A word of explanation is in order about the concept of Continental legal theory. The primary exemplar of this field is the theory of German criminal law as it developed in the early twentieth century and came into full bloom in the post–World War II period. The German literature of criminal law has been influential in the Spanish-speaking world, in Italy, in Eastern Europe, and in the Far East, especially in Japan and Korea. It is the primary alternative to the common law model of analysis, and most of our citations in this chapter are to German literature.

2.   TWO DIMENSIONS OF NECESSITY

So far as common law commentators have broached this classic conundrum, they have argued in the language of necessity.[4] Yet the plausibility of such claims of necessity may well turn on an internal tension in the theory of the defense. When common law lawyers speak of acting under necessity, it is never clear whether they are invoking a theory of justification or of excuse. As has been clear in German theory since James Goldschmidt's insightful article in 1913,[5] the rubric of necessity (*Notstand*) encompasses both theories of acquittal. When used as a justification, the theory of necessity requires a balancing of competing interests and a judgment that it is right, proper, and lawful to favor one interest over another.

When used as an excuse, the defense focuses not on the propriety of the act but on pressures compelling the defendant to violate the law. In the classic nineteenth-century case of *Regina v. Dudley and Stevens*, three starving shipwrecked sailors killed and cannibalized a fourth. The English court convicted them of murder in expectation that the Crown would commute their sentence, which in fact happened.[6] If the case had been tried in a German court, the shipwrecked sailors might well have been acquitted on grounds of necessity. A German court might have concluded that they killed the lad Parker and fed on his flesh in order, in the language of the German Code, "to avoid an imminent, otherwise unavoidable risk" of starvation.[7] But because this defense is one of excuse and not of justification, the killing would not be considered lawful and right. Nor could it be justified in a deontological ethical system that rejects the idea of killing an innocent person for one's own benefit. But the pressures of hunger and the imminence of death might excuse their doing the wrong thing. The two facets of the defense correspond to radically different inferences about the defendant's character. When we justify a choice between evils, we applaud the defendant's judgment in choosing the superior value. His decision reflects well on his character. In contrast, when we excuse conduct as necessitated by overwhelming pressure, we reject the suggestion that the defendant's decision tells us what kind of person he is. We attribute his decision not to his character, but to circumstances in which the human thing to do is to succumb to pressure.

How can one justify killing the psychotic aggressor under a theory of necessity? We would have to assume that sacrificing the life of the assailant yields a greater expected value, yet the most that can be gained from the killing is the saving of one's life. If it is life against life, it is hard to see why we should say that it is right and proper for one person to live and the other to die.[8]

The fact is that, in the case of the psychotic aggressor, we are inclined to favor an acquittal even if the loss to the aggressor is greater than the gain to the defendant. Indeed, as the problem is stated, this is the case. All the defending party knows is that there is a possibility of death if he does not resist. To fend off this possibility, he chooses certain death for the aggressor. When probability factors are included in assessing the competing interests, it is clear the defendant engages in conduct with a higher expected loss (certain death) than expected gain (a probability of death). We could decrease the threat to the defendant without altering our intuitive judgment about the desirability of an acquittal. Would it make any difference if the defendant were threatened with loss of limb, rape, or castration? One would think not. As the problem is treated in the literature, it is assumed that justice would require acquittal in these cases as well.

It is possible that those commentators who looked at the problem as one of necessity thought that the life of the insane aggressor is worth less than the life of the defendant who is standing his ground. One finds analogies between psychotic aggressors and attacks by wild animals.[9] If one thinks of the psychotic aggressor as subhuman, one might be able to justify the defensive killing as an act preserving the greater value. This is an intriguing if startling approach, but one that is apparently inadequate.

Among its other defects, it fails to account for the case of temporary psychosis. If the aggressor is a brilliant but temporarily deranged scientist, it would seem rather odd to say that his life is worth less than that of his victim, who for all we know might be a social pariah.[10]

More fundamentally, this approach violates a basic premise of the criminal law that individuals ought to be judged by what they do, not by their social status or general moral worth. It is true that in assessing culpability we make a judgment about the defendant's character, but that judgment is limited to his character as manifested in the commission of the proscribed act. For purposes of sentencing one might wish to engage in a free-ranging inquiry about the defendant's background and dangerousness, but those issues have traditionally never borne on the analysis of liability. Those who solve the problem of the psychotic aggressor by depreciating the life of the insane depart from the basic premise of equality before the law; they forsake the principle that it is conduct rather than status that determines criminal liability.

Depreciating the status of the psychotic reminds one of the moral devaluation of "rogue states" in international law. If one side is morally superior to the other, it is not particularly difficult to justify actions as furthering "the greater good." Yet international law is supposed to be committed to the equality of all states, just as the principle of equality of the well and the sick prevails in domestic law.

The insane are not a breed apart, not lunatics under the permanent control of the moon (as once believed), but mentally ill humans who can, in principle, return to a normal state of responsibility. Insanity is not a condition that removes the person from the scope of the law for all cases and under all situations. The law addresses them, too. They violate the law when they kill, but they do so under conditions that deny responsibility for a particular act under a particular set of influences.

The necessity of a theory of excuses derives from this moral aspiration of the law: If the norms of the law address the insane, then there must be some available account of why the psychotic aggressor is not responsible for his aggression. The theory of excuses comes on the scene to fill this gap, to explain how individuals can violate norms and yet not be held accountable.

Excuses are a relatively recent development in the history of legal thought. In the eighteenth century even the most advanced thinkers, such as Kant, sought to solve the problem of responsibility not by applying excusing conditions but by limiting the scope of the relevant norms. Or at least this is one way to read Kant's treatment of the problem. In an oft-cited passage in his *Legal Theory*, Kant poses the case of the ship-wrecked sailor who, lost at sea, saves his life by pushing another person off the only available plank floating amid the wreckage. Our intuitions tell us that we should not punish this person, who acts to save his own life. But why? Kant writes:

> There can be no penal law that would assign the death penalty
> to someone in a shipwreck who, in order to save his own life,
> shoves another, whose life is equally in danger, off a plank on
> which he has saved himself. For the punishment threatened by
> the law could not be greater than the loss of his own life. A penal
> law of this sort could not have the effect intended, since a threat
> of an evil that is still uncertain (death by judicial verdict) cannot
> outweigh the fear of an evil that is certain (drowning). Hence the
> deed of saving one's life by violence is not to be judged inculpable
> (*inculpabile*) but only unpunishable (*impunible*), and by a strange
> confusion jurists take this subjective immunity to be objective
> immunity (conformity with law).[11]

This passage lends itself to alternative interpretations. Either the statutory norm prohibiting homicide does not address the shipwrecked sailor if it cannot influence his behavior, or the rationale for not punishing the sailor's life-saving act is the exculpatory effect of necessity. Kant's views hover between these two positions. Even if the statutory norm does not apply to the shipwrecked sailor, his act is still a wrong in violation of general principles of law. This is the point of the last sentence, which stresses that the act is contrary to *Recht,* or law as principle. It is a wrong because it violates the general norm against interfering with the rights of others, and the first occupier of the plank presumably has a right to stay there.

Though the life-saving act of the shipwrecked sailor is wrong, the actor enjoys a subjective or personal immunity from punishment. Thus the life-threatening situation becomes a basis for excusing his conduct in the same way that the inner compulsion of the psychotic aggressor renders his conduct wrongful but excused. Both acts occur under overwhelming pressure. In one case, it is the imperative of staying alive that drives the sailor to aggress against the first occupier of the plank. In the other case, it is the distorting effect of mental illness that commands the act of aggression. We see, then, that there are two types of excuse: one based on external and the other on internal circumstances. In neither case can one say that the aggressor would choose to do the act if the situation were normal. The aggressor's true personality or character remains concealed behind the overlay of circumstances that compel the action.

Linking the theory of excuses to the actor's character raises thorny issues here, but it is not clear that we need to examine this linkage.[12] It should be enough to stress the involuntary nature of excused aggression. We cannot blame either the psychotic aggressor or the shipwrecked sailor because they are acting in extremis. They are not freely choosing to commit aggression. Yet the very language we are using—the vocabulary of voluntariness and free choice—troubles some theorists, who are skeptical about whether action is ever truly free and whether there is such a thing as free will. The skeptics, it seems, are inclined to ground the theory of excuses in their inferences to the actor's character from wrongful action. They think that by relying on the notion of character they can justify criminal punishment, without relying on what they consider dubious assumptions about free will.[13]

The tentative conclusion at this stage of the argument is that insanity and personal necessity represent parallel excusing conditions that negate the voluntariness—and thus the actor's personal responsibility—of a wrongful act. In Continental legal theory, certain features follow from the theory of excuses. If the actor is excused, his act remains wrongful (or unlawful), and that bears two systemic characteristics: everyone is allowed to use defensive force against a wrongful (and therefore unlawful) attack, and no one is entitled to aid a wrongful attack without becoming liable for the consequences.

It appears that neither dimension of necessity is adequate to solve the problem of the psychotic aggressor. The theory of justification requires the dubious assumption that the life of a psychotic is worth less than the life of the defendant. The theory of excuse yields unacceptable results with respect to the position of third parties who might intervene. The reason is that the theory of necessity ignores one important feature of the case: the fact that one of the parties is an aggressor and the other is holding his ground. That is precisely the feature that is critical in the theory of self-defense, and therefore we turn to that cluster of theories in our search for an adequate rationale for acquitting the slayer of a psychotic aggressor.

3. LEGITIMATE DEFENSE AGAINST A PSYCHOTIC AGGRESSOR

In discussing the *Goetz* case in chapter 2, we outlined four distinct approaches to the theory of self-defense: self-defense as punishment, as an excuse, as the vindication of autonomy, and as a variation of lesser evils. We now look at these arguments in greater detail as they apply to the problem of the psychotic aggressor. Because the psychotic would have a good defense by reason of insanity, we leave aside the first possibility of self-defense as a form of punishment, and are left then with three remaining models of defense that interact in different ways in different legal systems. We shall apply these models to the problem of the psychotic aggressor and then raise a new question: whether the model of the defense has an impact on the rule of proportionality.

The first model represents a theory of excuse. It is limited to those interests, such as life and bodily security, that individuals feel compelled to protect, and it typically requires the defendant to "retreat to the wall" before resorting to deadly force. The theory of the defense runs parallel to the excuse of necessity. The critical feature of both defenses is that circumstances have *compelled* the defendant to resort to deadly force.

This rationale is well expressed in the common law defense of se defendendo.[14] According to common law authorities like Coke, Hawkins, and Foster,[15] the essence of se defendendo is the "inevitable necessity" of killing to save one's own life. This provides a rationale for acquitting the party who defends himself against the psychotic aggressor, but it suffers from the same shortcoming as the theory of necessity as an excuse: it fails to generate a defense for third persons who choose freely to intervene on his behalf.[16]

The other two models of self-defense are both theories of justification, and thus they could potentially provide rationales for acquitting both the victim of the psychotic aggressor and the stranger who comes to his rescue. What is it about the aggression that prompts us to think that both the victim and the third party ought to be able to kill the psychotic aggressor? The approach of Continental legal theory is that the defender has a right to the integrity and autonomy of his body, and therefore he has a right to repel encroachments upon his sphere of rights. Let us refer to this theory, in the parlance of German law, as *necessary defense*. The essential condition for necessary defense is a particular kind of aggression against an innocent agent; not all forms of aggression encroach upon autonomy. If the aggression is justified by, say, an arrest warrant, then even an innocent party under arrest would not have the right to resist. The criterion for aggression that may be resisted is unlawful (or wrongful) conduct, where it is clearly understood that conduct that is merely excused—for example, the attack of a shipwrecked sailor over the plank—is still wrongful. The focus is not on the culpability of the aggressor but on the autonomy of the innocent agent. It is assumed that the innocent defender has a right to prevent any encroachment upon his autonomy. As German scholars have put it, the Right should never give way to Wrong.[17]

Necessary defense is the dominant theory in the entire German sphere of influence, and it also found expression in early common law theory. Recall Sir Edward Coke's insistence that "a man shall never give way to a thief, etc., neither shall he forfeit any thing."[18] John Locke supported the same theory of an absolute right to protect one's liberty and rights from encroachment by aggressors.[19] (A graphic illustration of this theory is found in Tom Wolfe's novel *Bonfire of the Vanities*.[20] The protagonist, a white, upper-class professional, is put in a holding cell with a group of men much tougher and prison-hardened than he; one steps over and asks the newcomer to give him the cup of coffee he is drinking, and the protagonist realizes at that moment that if he surrenders the cup of coffee, he will become the slave of the covert aggressor.)

There is a tendency in this literature to think of the aggressor as someone who has placed himself beyond the law. Locke, for example, speaks of the aggressor's being in a "state of war" with the defender,[21] as if the aggression breaches an implicit contract among autonomous agents, according to which each person or country is bound to respect the living space of all others. The intrusion upon someone's space triggers a justified response.

The same doctrine finds expression in the rhetoric of the American Revolution. The slogan "Don't tread on me" expresses the claim that treading on someone else in itself entails a justified response. It is irrelevant whether the country or person so treading is culpable for his deeds. When at war, one is concerned only about the enemy's aggression—not about his possible excuses. The roots of the right of defense are not in the culpability of the aggressor, but in the autonomy of the defender.

It is curious, however, that Article 51 of the United Nations Charter permits self-defense against any "armed attack," without stipulating that the attack be unlawful or wrongful. Suppose that Israel's strike against Egyptian airfields in the Six-Day War was justified as a response to an imminent attack by Egypt. Would Egypt not be justified in responding to the Israeli armed attack? There is no way to answer this question. The literature of international law is insensitive to this distinction. Not even the Rome Statute of 1998 recognizes the relevance of the distinction between justification and excuse in formulating the defenses to the crimes within the jurisdiction of the International Criminal Court.[22] Thus, when we discuss these questions in the international arena, we work from a blank slate. Although the theory of justification is highly relevant to the law of war, it is not so clear, as we shall see, whether the theory of excuses ever affects the analysis of self-defense under Article 51.

Necessary defense, as we have described its principles here, generates a paradoxical view of aggression. It treats the aggressor as both a participant in the legal system and an externalized threat. The aggressor's wrong-doing matters, but his culpability does not, and if tried for the crime, he would receive the benefit of all available excuses, including insanity, duress, mistake, and involuntary intoxication. But none of this is available to the aggressor in the face-to-face confrontation with the defender. The defender does not have to pull his punches just because the aggressor is insane or drunk or mistaken. In such a confrontation on the street, the aggressor loses the protections that he would normally get during a trial, that is, the right to plead excuses such as insanity or duress. It is almost as though the aggressor no longer counts as a person. Thus the aggressor is at once inside and outside the legal community, simultaneously a comrade and an outlaw. This is a paradox generated by the law's treatment of aggressors.

This ambiguous status is reflected in conventional legal expressions. In the German idiom, each instance of self-defense represents a conflict between *Recht* (Right) and *Unrecht* (Wrong). The victim stands for the Right, the aggressor for the Wrong, as though each conflict were a drama between good and evil. Another central concept is that of the legal order (*die Rechtsordnung*), which in the German view is threatened by every breach of the law. Thus the autonomy of the defender is identified with the idea of Law itself, and every act of self-defense is a defense of the basic structure of legal relationships. This explains why necessary defense is available to the defender as well as to third parties. This conceptual framework shunts the aggressor to the fringes of the legal system. Though he may be psychotic and morally innocent, he becomes an enemy of the Law itself.

The alternative, common law theory of self-defense is a variation on the justification of necessity; it originates from the need to balance the interests of the aggressor against the interests of the victim.[23] Yet this model of self-defense as a lesser evil permits one to kill in the name of an interest less valuable than life by adding another factor to the balancing process—typically, the culpability of the aggressor. The aggressor's culpability becomes the rationale for diminishing his interests relative to those of the victim. The common law would argue that one simply cannot balance the life of a culpable aggressor against the life of an innocent victim on the assumption that the two combatants are equally situated. The one who chooses to start the fight is held to be entitled to less protection than the innocent victim. But the problem is how significant the factor of culpable aggression ought to be in diminishing the interests of the aggressor. This is the critical variable in assessing whether self-defense ought to be available against rape, castration, maiming, and theft as well as against homicide. The more significantly one regards the culpability of the aggressor, the less significant the victim's interest has to be in order for the victim to have the right to use deadly force, if necessary, to repel the attack. The underlying premise is that if someone culpably endangers the interest of another, his interests are less worthy of protection than those of the innocent victim.[24]

Self-defense as the lesser evil has found support in Germany,[25] England, and the United States, but it has become dominant only in the common law tradition. Blackstone made it clear that the criterion that determined punishment (namely, the gravity of the criminal attack) also prescribed the contours of self-defense. No act "may be prevented by death," he wrote, "unless the same, if committed, would also be punished by death."[26] Less serious attacks, such as petty thefts, were not capital offenses, and thus they could not be resisted with deadly force. Balancing competing interests is the essence of this version of self-defense, and the rule of reasonableness (or proportionality) naturally adheres to any system that, like the common law, is committed to the theory of self-defense as the lesser evil.[27]

The pinion of the balancing process is the aggressor's culpability. The aggressor's culpability depreciates his interests in the balancing process and thus distinguishes self-defense as the lesser evil from the justification of necessity. Yet with culpability as its pinion, this theory of self-defense cannot be geared to solve the problem of the psychotic aggressor. By definition, the psychotic aggressor is not culpable, and thus the Blackstonian common law approach fails to explain why his interests should be worth less than those of the victim. One might try to argue that the mere aggression, whether culpable or not, provides a basis for diminishing the interests of the aggressor in the balancing process. Yet this argument does not survive critical examination. Why should someone's interest be diminished by virtue of aggressive propensities for which he is not morally to blame? One can understand diminishing an aggressor's interest if he is to blame for the encounter, but it is hard to see why he should be worth less merely because his body is the locus of dangerous propensities. Self-defense as the lesser evil may provide a perfectly sound rationale for self-defense in the typical case, but it fails to explain our intuition in a case in which the aggressor's conduct is excused by reason of insanity or other acknowledged excusing conditions.

4. PROPORTIONALITY IN DOMESTIC LAW: KILLING AN APPLE THIEF

As one may imagine, the doctrinal lens of necessary defense, as recognized in the Continental tradition, filters out the shades and nuances of each case and transforms all situations into black and white. The only question left is whether the aggressor has intruded upon the defender's sphere of autonomy; questions of degree are suppressed. This absolutist perspective, however, proves hostile to the rule of proportionality, as confirmed by the German Supreme Court in 1920 in the classic case in which an orchard owner shot and killed a thief running away with fruit from his trees.[28] He claimed that he shot in self-defense—namely, in defense of his property. He was found not guilty; the prosecutor appealed, and the Supreme Court affirmed. The court recalled the basic premise of the Kantian theory of defensive force: Right need never yield to Wrong. As Kant formulated it, the foundation of Right (or the Law) is the principle of external freedom. Each person is entitled to a maximum degree of freedom compatible with a like freedom for others.[29] Thus any intrusion upon a person's sphere of rights generates a subsidiary right to use whatever force is necessary to repel the aggressor and restore the defender's autonomy.

For generations, German scholars and their followers around the world have been anxious about this problem of shooting petty thieves. Of the many approaches in the literature and the case law, one draws on the first word in the definition of necessary defense in § 32—"appropriate"— which provides, "Necessary defense is that defense which is *appropriate* to avert an imminent unlawful attack against oneself or another" (emphasis added). Thus the scholars and judges would cut back the use of necessary defense on the ground that it is not "appropriate" under certain circumstances. This approach is somewhat less than principled, but at least it is grounded in the statutory language.[30]

Another German attempt to make the distinction resembles the argument in the common law, which has progressively revised the categories for using defensive force by distinguishing first between felonies and misdemeanors, and thereafter between "atrocious" felonies and less serious felonies.[31] A few German writers have tried the same approach. Friedrich Oetker argued in 1930 that attacks on minor interests should be regarded as mischief (*Unfug*) rather than wrongdoing (*Unrecht*) and that they should not trigger the full response permitted in a case of self-defense.[32] Eberhard Schmidhäuser revived this effort by claiming that petty attacks, those usually tolerated by society at large, do not represent violations of the Legal Order.[33] His example is the violation of an antinoise regulation by a group of motorcyclists. The only way to suppress the noise might be to shoot them—a patently absurd result, but one suggested by the Kantian principle of absolute rights. These kinds of violations would not generate a right of self-defense, for they are implicitly tolerated by the community as a whole. This is an intriguing idea, if only because it reveals the sophistication required to accomplish what is simply assumed in the common law tradition.

The alternative to thinking about appropriate defense is to invoke a doctrine of private law that used to restrict the exercise of seemingly absolute rights. Continental civil theories originated the concept of *abus de droit* ("abuse of rights") as a way of restricting the scope of certain rights in the law of contract and property.[34] This particularly Continental invention functions very much as an analogue to the rule of reasonableness in the common law. The difference is that reasonableness already figures in the statute as a limitation: "A person may use a reasonable amount of force in self-defense if he or she reasonably believes that he or she is about to be attacked." The German approach, which is followed in many countries in the Continental sphere, is to state an absolute rule in the statute—that is, a person may use all the force that is necessary to avert the attack—and then qualify the rule with a principle of abuse of rights when the use of force seems egregious and excessive.

Of course, this rule of limitation is hardly precise. Some scholars object that these vague boundaries create a trap for the unwary. We could once again appeal to Holmes's words: "Detached reflection cannot be demanded in the presence of an uplifted knife."[35] It is unfair to make the defender bear the risk of using excessive force (and going to prison) without providing clear guidelines specifying when he must forgo his defense and when he need not. If the use of force under § 32 is to be limited, then there must be unmistakable criteria for the kinds of cases where self-restraint is mandated. Good examples would be attacks by children or the obviously insane or the precise situation that arose in the 1920 apple thief case: When someone is escaping with stolen property, the owner must forgo deadly force and rely instead on available legal remedies. In all these cases, the defender would abuse his rights by shooting to kill.

One of the leading cases favoring the doctrine of abuse of rights is a 1963 Bavarian decision in which the defendant was convicted of attempted coercion for driving toward a pedestrian and threatening to run her down.[36] The woman was standing in the single parking space available in a lot where the defendant wanted to park. The victim stated that she was holding the spot for someone else, whereupon the defendant drove toward her. His argument was that the victim's behavior was illegal and that he was exercising self-defense. Fascinatingly, the Bavarian appellate court agreed. But, the court added, in this situation the defendant had abused his right of defense, for "the harm inflicted was disproportionate to that threatened by the attack" of standing in the parking space.

There would seem to be many reasons not to take this case as a precedent requiring tolerance of thefts or petty physical injuries when the only alternative is death to the assailant. First, this is a case on the fringes of self-defense. It is dubious whether one should regard occupying a parking space as an attack akin to physical assault. Second, it seems obvious that the defendant could have used lesser means to remove the woman from the space. The force was excessive even apart from the issue of proportionality. This case was once widely cited to support the growing scholarly consensus in favor of applying the doctrine of abuse of rights (*Rechtsmissbrauch*) to impose limits on the right of self-defense. This position has now become so well accepted that an extrastatutory limitation of proportionality on self-defense seems to be taken for granted.

Though this doctrine is gaining ground, it poses a significant difficulty. The principle of nulla poena sine lege (no punishment without prior legislative enactment) has constitutional status in Germany.[37] Yet the judicially legislated right of self-defense may violate this principle. Applying the principle of abuse of rights to cases of self-defense creates a new class of punishable acts that would be covered by the traditional right—but not the qualified right—of self-defense. Accordingly, these acts are rendered punishable by judicial decree, but thus imposing liability violates the rule of nulla poena sine lege.[38] This does not inhibit legislative reform, but it does suggest that circumscribing a defense by judicial decree is a dubious practice.

This argument has hardly won the day. German courts have gone even further and ruled, in a series of decisions, that East German border guards could be held liable for shooting at citizens trying to escape to the West. They were arguably justified under an East German statute that authorized their using deadly force to prevent major crimes.[39] Yet, after unification, the German courts reasoned that a statutory justification that violated higher principles of law could be disregarded in criminal proceedings against the guards. In effect, the German courts created a new category of criminal offense: shooting at escaping citizens in a way that previously would not have been punished under East German law. If this is an acceptable deviation from the principle of nulla poena sine lege, then curtailing the scope of self-defense is a minor adjustment.

The European Convention on Human Rights provides an additional source of limitation of the absolute right of self-defense recognized in the German Criminal Code. Under its recognition of the right to life in Article 2, the ECHR limits the role of self-defense to the protection of personal interests, thus excluding property, however valuable. Initially, the German courts tried to avoid this strongly worded limitation on necessary force to defend interests of property by arguing that the Convention applied only to interstate relations and had no bearing in actual cases within the domestic legal order.[40] The Court of Human Rights in Strasbourg now intervenes so often and aggressively in domestic cases of European criminal justice that this argument has lost its traction, and in the future the German courts will have to pay more attention to the terms of the Convention as limitations on their domestic practice.

Yet this European principle would hardly make much sense if applied at the level of international law. Imagine a case of bloodless invasion. Argentine troops take over the Falkland Islands without spilling, or even threatening to spill, a drop of blood. Would this not constitute a case of aggression that could be resisted by defensive force under Article 51 of the Charter? It seems clear that England has the right to defend its territory per se, and not only the people who live on that territory. Why, then, should individuals forfeit their right of defensive force when only property issues are at stake?

In fact, with respect to the occupation of land, there is a plausible distinction between the domestic and international arenas. Law enforcement in the domestic sphere is typically sufficient to regain land lost to unlawful occupiers, but if the Argentines were to gain control of the Falklands, it is not clear how the United Kingdom could invoke a legal process to regain its rights.

5. JUSTIFICATION AND EXCUSE IN INTERNATIONAL CRIMINAL LAW: ABSOLUTE TABOOS

One of the great puzzles raised by the Rome Statute is determining the range of conduct and crimes to which these defensive claims should apply. They do not apply to waging war as such. As we saw in chapter 1, the law of war does not need a doctrine of self-defense to justify soldiers killing other soldiers. The law creates a conceptual space in which combatants are reciprocally exposed to the risk of being killed. Their killing of each other, in the ordinary course of fighting, is exempt from the definition of war crimes. Therefore no "ground for excluding responsibility" is necessary to explain why they are not guilty.

Not all domestic criminal lawyers pay attention to the distinction between justification and excuse. Despite the recent revival of theoretical work on criminal law, most Canadian, English, Australian, and French lawyers lump the two together under the general category of "defenses." This explains why the Rome Statute, shaped by the English- and French-speaking world, appears to lump all defensive considerations together in Article 31 as "grounds for excluding criminal liability." The Article contains two grounds that are essentially excuses based on involuntary behavior (insanity and involuntary intoxication), one that is essentially a claim of justification (necessary defense) and one that is mixed (duress).[41] For our purposes it is worth looking more closely at self-defense and duress under the Rome Statute.

The provision on self-defense in the Rome Statute is drafted in the spirit of the Continental doctrine of necessary defense.[42] It covers all three relevant categories: defense of self, defense of others, and defense of property. The scope of permissible force in defense of self and of others is exactly the same: against "an imminent and unlawful use of force in a manner proportionate to the degree of danger to the person or the other person." This language conforms to the general requirements of self-defense in all domestic legal systems, except that it fails to mention the element of necessity. In the defense of property, the defender is entitled in the course of war to protect "property which is essential for the survival of the person or another person or property which is essential for accomplishing a military mission."[43] But the provision does not say how much force is permissible to defend these essential forms of property. It could be read in several ways: either as proportionate force or as any amount of force necessary. On the whole, however, this provision is in line with the general trends of criminal legislation in the world as a whole.

By contrast, the provision called "duress" is an unfortunate mixture of both excuse and justification. The term duress usually connotes a coerced response to a serious threat of harm, and under the Rome Statute, the defense is available when action is "caused" by any "threat of imminent death or of continuing or imminent serious bodily harm against that person or another person." The threat need not be unlawful. It could derive from human sources or from natural causes. It might even consist in a routine threat of death in warfare.[44] In responding to the threat, the person must act "necessarily and reasonably." Thus far, except for the omission of the word "unlawful," the language of the Rome Statute runs parallel to the definition of duress in the Model Penal Code and of "personal necessity" in the German Penal Code.[45] But the concluding phrase of the provision imposes a limitation not found in the code provisions on excusable conduct: "provided that the person does not intend to cause a greater harm than the one sought to be avoided."[46] The implication is that the provision is not applicable in homicide cases like the situation in the *Erdemović* case, tried before the International Criminal Tribunal for the Former Yugoslavia. In that case, the defendant, a Serbian soldier, participated in the slaughter of Bosnian Muslims after his commander threatened to kill him and his family.[47] Although Judge Cassese argued vigorously that duress could apply even in a war crime, the majority of the court held that duress can never be a complete justification for war crimes or crimes against humanity when the victims are innocent human beings.[48] Indeed, the language of balancing the harm committed against the harm avoided is not typically found in the definition of duress, but rather in claims of justification based on avoiding the lesser evil.

This infelicitous language compresses two major categories of defensible conduct that stand separately under most criminal codes. First, it does not include the full range of excusable conduct "caused" by overwhelming threats. In principle there is no reason why the excuse of duress should not apply to cases of coerced conduct, even if the harm done is greater than the harm avoided. This is widely accepted in Continental as well as U.S. jurisprudence. Further, the provision lops off cases of justifiable conduct—here the harm avoided exceeds the harm done—if that conduct is not "caused" by the external threat. The language of causation invokes images of coercion, but justifiable necessity might also be freely chosen as the right thing to do under the circumstances.

To correct this truncated combination of duress as an excuse and necessity as a justification, the International Criminal Court will have to improvise new defenses when the occasion warrants it. Fortunately, the Rome Statute permits the court to recognize new grounds for excluding criminal responsibility when they are available in the "general principles of law derived from national laws of legal systems of the world," and both duress as an excuse and lesser evils as a justification meet this test.[49]

One of the great puzzles raised by the Rome Statute is determining the range of the conduct and crimes to which these defensive claims should apply. They do not apply to waging war as such. As we shall see in chapter 8, the law of war does not need a doctrine of self-defense to justify soldiers' killing other soldiers. The law creates a conceptual space in which combatants are reciprocally exposed to the risk of being killed. Their killing in the ordinary course of fighting is exempt from the regular prohibition against killing, and therefore no "ground for excluding responsibility" is necessary to explain why they are not guilty.[50]

The range of Article 31 is extremely limited. For example, we cannot imagine that the crime of genocide could either be excused or justified. Genocide in self-defense or under duress seems a bit far-fetched. If a case of self-defense or duress or necessity could even be contemplated in this case, the purposes of the action would be beneficial, and no longer subject to description as "an intent to destroy the group as such." And it seems equally implausible that participants in genocide could rely on claims of insanity or involuntary intoxication. Perhaps one or two participants might be so affected in their mental capacity, but there would still be many in the group who would retain their capacity to act responsibly.

Crimes against humanity seem equally beyond the realm of the arguments detailed in Article 31. How would one seek to justify the crimes of extermination, enslavement, or torture on grounds of self-defense, necessity, or even the peculiar form of duress defined in the statute? These, as well as other offenses listed in the catalogue of crimes against humanity, are absolute taboos. Any attempt to justify torture on grounds of lesser evil would generate the kind of controversy we have witnessed in the United States in recent years over the "torture memos" written to justify the behavior of the U.S. troops in Abu Ghraib.[51] The same is true of "enforced disappearance of persons."[52] Should we give a brutal tyrant his day in court with the claim that kidnapping and murdering suspects was necessary for the defense of the country? Nor is it easy to imagine the systematic or widespread commission of these offenses as excused on grounds of insanity or involuntary intoxication. Groups might go temporarily crazy or get temporarily drunk, but they are not likely to act long enough under these conditions for their conduct to meet the criteria of collective crimes against humanity.[53] When it comes to the many crimes that egregiously violate human dignity, the very thought of a justification is an insult to the victims.

The prohibitions against sex crimes are also absolute in nature. Think of rape and sexual slavery.[54] No one could justify these actions on grounds of self-defense or lesser evil. And insofar as we think of them as collective actions, they are not subject to excuse either. But as individual war crimes they might be subject to claims under Article 31(1)(a) and (b) by analogy to sex crimes committed in domestic law.

But there are some crimes on the long list subject to the jurisdiction of the International Criminal Court for which the argument of justification might be thinkable, even plausible. All offenses of destroying property, we would think, are subject to possible justification. And others might be justifiable where necessitated by the exigencies of war. Forcible transfer or deportation would be justifiable if undertaken to protect the population from atomic radiation or some other disastrous by-product of war.

There is no way to know, at the outset of an investigation or prosecution, whether or not the provisions of Article 31 apply. This problem is radically undertheorized. The commentators do not address the issue.[55] The readers of the Rome Statute must have the good sense to distinguish between offenses that are meant to be absolute taboos and offenses that function as most crimes do in domestic criminal law: criminal under normal circumstances but subject to justification in abnormal cases.

In the final analysis, then, the impact of Article 31 of the Rome Statute is limited indeed. We can imagine some cases of attacks by civilians that must be repelled by the use of military force, taking a toll both on lives and property; there would be no war crimes in these limited cases. There might also be a few other incidents in which soldiers executed commands under duress, and they would be excused if the harm was not too great. However limited its range, Article 31 is a good first step in refining international law by applying justification and excuse to the larger issues of justifiably going to war or coping with the risk of terror at home by declaring a metaphoric "war on terror."

6. EXCUSES IN INTERNATIONAL LAW

Let us shift now from jus in bello to jus ad bellum, from incidents of prosecutable criminal conduct to the conduct of warfare in the international law binding on states. The only case of lawful war as we know it is as an exercise of legitimate defense, as a necessary and proportionate response to an imminent unlawful attack by another state or an equivalent substate organization against one's own nation or another. The very existence of other defenses in Article 31—other "ground[s] for excluding criminal responsibility"—prompts us to ask whether these other grounds could find application on the international level. What about the possibility of duress as defined in the Rome Statute, or lesser evils as we might understand them under the creative potential of the International Criminal Court?[56]

If populations led by their armies were forced to move into other countries to avoid a natural disaster, the receiving country would not likely regard the newcomers as invaders; presumably, they would be treated as temporary refugees. Of course, the receiving country could close its borders, and if that happened, the incoming population might use weapons to clear the way and generate living space for themselves. If their incursion was by force, they would clearly be in violation of Article 2(4) of the United Nations Charter, and their arguments of duress and necessity would fall on deaf ears. This thought experiment convincingly illustrates that these other claims of excuse recognized in domestic law would find little application in international law.

National autonomy seems to be a stronger value than personal autonomy in Western legal systems today. No state can make the decision that it should invade another state by reasoning that on balance the invasion is better for the international community as a whole. On the international level, our communal sense of shared interests is not yet strong enough to overcome traditional principles of national independence. This issue will be discussed in greater detail in the next chapter on humanitarian intervention.

The likelihood of invoking excuses in international conflict is minimal. As we mentioned in chapter 3, international law once viewed self-defense as an excuse based on self-preservation, but international lawyers now exclusively treat the concept as a justification. War is based on collective action (a theme we explore more thoroughly in the final chapter), and most excuses require a mental element hard to conceptualize in the case of nations.

Consider again the case of the psychotic aggressor, but this time apply it to the international system. Assume that a psychotic dictator, without popular support, launches an attack against a neighboring state. The dictator is clearly acting irrationally and has a mental disease, say, megalomania and narcissism rising to the level of a psychotic misperception of his place in world history. The dictator also has complete control over his country's military, so he has the ability to implement his psychotic plans. Let us also assume that the inhabitants are afraid to resist because in the past the dictator has summarily executed both deserters and political opponents. He shows no mercy at all.

The attack is therefore launched against the neighboring state, and it is clearly unlawful aggression under international law. The target has the right to resist the unlawful aggression, but how do we evaluate the international culpability of the dictator's citizens? We would have no problem holding them responsible for the dictator's actions, at least as long as the dictator remains in power. But what if the people overthrow their dictator and renounce his past actions? There is a basic assumption that a state is responsible for its past actions even after a change in government. For example, past debts are assumed by a new government, and this is usually the price of admission to the world community. It is therefore a pragmatic requirement under international relations that a state treat itself as being continuous with its predecessors, even if it has changed heads of state or its entire political structure.

Nonetheless, one might imagine the inhabitants of this newly liberated state appealing to the international community that their conduct ought to be excused. Their nation was hijacked, they might say, by a psychotic dictator. The nation as a whole ought to be excused for the psychosis of its leader. Under domestic law, the psychosis would be accepted as a good defense. How should we treat the excuse under international law? Of course, the excused aggression is still unlawful and the target of the invasion should be allowed to resist with force. Legitimate defense is always allowed against an unlawful attack regardless of whether the original attack is excused or not. The aggression was not, after all, *justified*. But the international system may wish to *excuse* the aggression and withhold punitive sanctions against the rehabilitated nation, even though it was the source of the aggression. The international community may feel that it would not be fair to burden its citizens with the crushing weight of war reparations.

Should we use the language of excuse to describe this situation? We are not so sure, although the possibility is provocative. There is, however, one claim of excuse that nations might be able to successfully invoke to defend themselves against charges of unlawful aggression: if they are reasonably mistaken about the possibility of an imminent attack, their response will subsequently be excusable. Suppose that the United States had credible and reasonable public evidence that Saddam Hussein possessed weapons of mass destruction, as well as a long-range missile system aimed at the United States. Under such a perception of danger, the invasion of Iraq could have been defended even if no weapons were subsequently found. In the domestic law of self-defense, a reasonable belief in an imminent attack will excuse the use of force, even against someone who is totally innocent. In Continental jurisdictions this is called *putative self-defense*. Some people argue that putative self-defense is also justified on these reasonable perceptions of danger, but such debates about classification are esoteric and not subject to easy resolution, and for our purposes, we need not decide whether putative self-defense is an excuse or a justification.[57] The salient point is that a reasonable mistake of danger can support the use of force in the international arena.

Most international lawyers treated the legitimacy of the Iraq invasion as if it were dependent on actually finding WMD. As far as the analogy to domestic law is concerned, this is false. It should have been enough that the evidence prior to the invasion was public, observable by all, credible, and sufficient to satisfy a reasonable person. Adlai Stevenson's presentation of photographs of Russian missiles installed in Cuba provided a benchmark for the kind of evidence that should have been produced. However, the evidence that Colin Powell revealed at the United Nations (pictures of unmarked trailers where weapons were supposedly being manufactured) fell far below that standard of reasonable proof. A full analysis of this "mistake of fact" excuse for the invasion of Iraq will be presented in chapter 7 on preventive war.[58]

Whether putative self-defense—reasonable mistake of fact as an excuse—should be accepted in international law remains an open question. Should there be a standard of the "reasonable nation," by analogy to the reasonable person, for judging whether mistaken perceptions of aggression are excusable? Should the prior relationship of, say, Iraq and Israel enter into the analysis, as the prior relationship of the parties would in domestic law? These are urgent issues, but the literature on international law has not yet begun to address them. It would be gratifying to see an awakening of interest in the foundations of such principles that are taken much too much for granted.

6

HUMANITARIAN INTERVENTION

An armed attack is justified under international law only if authorized by the United Nations Security Council under its Chapter VII powers or by inherent self-defense. But who gets this right? Is it nations, states, or "peoples" in the broadest sense of the word? In this chapter we explore the possibility that the inherent right belongs to nations, not just states admitted to the United Nations. Although there is a tendency in international law to do away with the concept of the nation entirely and to deal exclusively with states, we believe our view ultimately accords with the collective right of self-determination recognized by the UN Charter. If successful, this argument will have serious implications for our doctrine of legitimate defense: all nations, whether recognized as states or not, might have the right to engage in collective self-defense. Furthermore, the doctrine of legitimate defense would imply that other nations can come to their aid if the same conditions are met. All of these issues must now be critiqued and explored.

1. INTERVENING FOR THE SAKE OF HUMANITY

On April 6, 1994, a missile struck an airplane flying above Kigali and carrying the Rwandan president Juvénal Habyarimana, killing him as well as the president of Burundi. The assassination immediately reignited a bloody civil war that included one of the worst genocides of the twentieth century. Hutu extremists seized the opportunity to launch a genocidal campaign against Tutsi and moderate Hutus. But their weapons of choice were not gas chambers, as the Nazis had used, or futuristic nuclear weapons. Their weapons of choice were machetes.

Roving bands of Hutu militiamen went door-to-door looking for Tutsi and hacking them to death one at a time. Newspaper articles encouraged Hutu civilians to aid in this effort and kill Tutsis that lived in their villages. Radio programs broadcast the location of fleeing Tutsi and called on Hutu citizens to do their duty and kill them. The result was that about eight hundred thousand Tutsi and moderate Hutus were exterminated in a bloodbath. And the killers were not limited to a handful of army commanders and their soldiers. This was a decentralized genocide, carried out by tens of thousands of Hutu militia—and civilians—who picked up a gun, knife, or machete and killed their neighbors just because they were Tutsi.

The killing was not performed in secret. It was done openly and in public, and the world knew about it. Indeed, the world was given advance notice about it. The United Nations already had peacekeepers on the ground who tried to keep the simmering ethnic conflict under control. The troops were under the command of Canadian General Roméo Dallaire, who received intelligence reports from the field that Hutu extremists were stockpiling small weapons and planning something.[1] Dallaire dispatched an urgent report to UN headquarters in New York about the stockpiling and requested more troops to stop the coming attack. His report was ignored, although it is unclear whether Secretary General Kofi Annan saw the report and ignored it or whether the report got lost in the UN bureaucracy.[2] When the killing started, there was nothing Dallaire could do without more troops.

Dallaire never got more troops because the U.S. government was against a larger international military force in the region and had privately expressed its skepticism to other world leaders. Any attempt to launch a larger international force would have risked a U.S. veto on the Security Council. President Clinton had recently gotten stung by Operation Restore Hope, the U.S. military intervention in Somalia that resulted in the deaths of eighteen U.S. servicemen in 1993. The Somali government had collapsed and the country was being run by rival warlords and their militias. Relief supplies and food were being stolen by gangs and little of it was getting to Somali civilians, many of whom were getting shot and killed. Relief workers were being blocked. U.S. and UN troops were already in Somalia to restore order and pave the way for relief supplies, though little progress was being made.[3] Clinton then mobilized U.S. Army Rangers and Delta Force commandoes with orders to arrest the warlord Mohamed Farrah Aideed. But when a Blackhawk helicopter was shot down by Somali militia, the soldiers got bogged down in a vicious street fight. A Blackhawk pilot was taken prisoner, eighteen soldiers were killed, and newspapers around the world carried photographs of Somali militia dragging the bloody corpses of U.S. soldiers through the streets of Mogadishu. Though it was the Somali militia that killed the soldiers, it was these photographs that effectively killed Operation Restore Hope.

After Clinton got the pilot back and withdrew the troops from Somalia, he was wary of "restoring hope" anywhere else in the world. The genocide in Rwanda was clearly horrific, but Clinton was not interested in sending U.S. troops to stop it. In fact, administration officials were unwilling to even refer to events in Rwanda as "genocide" because they felt this might imply a duty to intervene. There was a feeling at the time that the word "genocide" was legally significant. The Genocide Convention, signed by the United States in the aftermath of World War II, imposed a duty on all signatories to "confirm that genocide, whether committed in time of peace or in time of war, is a crime under international law which they undertake to prevent and to punish." The relevant phrase here was the requirement that states must "undertake to prevent" genocide. Some scholars had suggested that not only would the Genocide Convention legitimate foreign intervention, but it might also *require* intervention as a matter of international law.[4] For a while, this argument was the darling of human rights activists, because it turned the tables on the question of humanitarian intervention. Not only was intervention permitted, but a state's failure to intervene might violate its international responsibilities. Because Clinton was not ready for intervention, he stayed away from any public statements that might suggest that the administration had concluded that the Hutu rampage was a genocide.

The prevailing wisdom about the Genocide Convention has since changed. Although it was a nice idea, no one really takes seriously anymore the idea that the Convention obligates signatories to launch military forces to prevent genocide. There may be a *moral* duty to launch military campaigns if a state has the capability to do so, but a moral duty is entirely different from a legal duty under international law. Although the Genocide Convention says that states must prevent genocide, this must be read in tandem with the general assumption of international law that military force is always exercised voluntarily. Even the Security Council when it authorizes military action does not require an individual state to participate.

Ironically, the Bush administration has been much more willing to use the word "genocide." Because the term is not considered to be a trigger for responsibilities under international law, the term is much less significant. The Bush administration has been much less reticent about calling the civil war in Darfur a genocide. There is substantial debate in the academic literature over whether the killing in Darfur is a genocide or "just" a crime against humanity and war crimes.[5] But the important point is that the Bush administration did not believe that calling Darfur a genocide required intervention.

The Bush administration has shown absolutely no interest in intervening in Darfur, much as Clinton showed no interest in intervening in Rwanda. The Darfur situation is considered by all involved to be a humanitarian crisis, but U.S. forces are otherwise occupied by the war against Al Qaeda and the Taliban in Afghanistan, the war in Iraq, and the ever broadening war on terror. The idea of sending troops across the globe for humanitarian reasons to countries where the United States has no strategic interests sounds almost quaint. But the real reason is that U.S. politicians have not forgotten the photos from Somalia, and the U.S. public needs a really good reason to accept their soldiers coming home in body bags. The image of the World Trade Center collapsing on 9/11 provided exactly this kind of reason. Not surprisingly, the image of a starving family, or a burning village in Darfur, does not.

The great exception to this story is Kosovo. The United States refused to intervene in Rwanda and the Sudan, but we bombed Serbia to force its troops from Kosovo. Milosevic had already fought two wars against Serbia's neighbors in pursuit of an ethnically homogeneous Greater Serbia. Militias rounded up men of fighting age and drove them to fields, where they were executed. Others were held in camps that were a shocking echo of the Holocaust. The fighting ended only when the Dayton Peace Accords were negotiated by Madeleine Albright and Richard Holbrooke and NATO bombed Milosevic and the Serbs to make clear that we would back up our negotiations with military force if necessary.

But just a few years later, Milosevic was back at his campaign of genocide and ethnic cleansing. He claimed that Kosovo historically "belonged" to the Serb "people," and they would rid their province of all Albanians. A military campaign was launched, but much of the terror was carried out by militia in civilian clothes, who went door-to-door raping women and killing Albanian men. A flood of refugees fled the villages in search of safety. This left them even more exposed to Serbian forces. Clinton was motivated to stop the genocide but had learned his lesson from Somalia. If intervention was going to be successful, it would have to follow three simple rules. First, the United States kept troops off the ground as much as possible and relied on air power instead; the use of ground forces was what got us in trouble in Somalia. So we bombed the Serbs instead of launching a ground invasion. Although this limited our options and resulted in collateral damage, at least we were not putting U.S. personnel at risk. Second, we went in with overwhelming firepower. The Somalia mission was carried out by a warship and one unit of Army Rangers and Delta Force commandoes. We would not make the same mistake twice. The war in Kosovo involved an overwhelming number of F-15s and F-16s launching sorties every few minutes and dropping an unprecedented number of bombs. Milosevic knew that we meant business. Third, the United States went in with an international force backed by NATO instead of an international force backed by the United Nations. Indeed, the Security Council never voted to authorize the Kosovo bombing. The member states of NATO, a mutual defense pact, did. The general in charge of the campaign, Wesley Clark, was the supreme allied commander of NATO Forces. The Russian government was more sympathetic to the Serbs than was the United States, and they indicated that they would probably veto any Security Council attempt to authorize military force to stop the genocide.

This last point raised an uncomfortable question of international law. Because in international law the use of force is allowed only in cases of self-defense against an armed attack or Security Council authorization, the bombing of Serbian forces may have been illegal.[6] Yet the intervention seemed like the right thing to do. There are only two approaches to resolving this tension: either construct a *legal* argument for humanitarian intervention or concede that the bombing was illegal but that morality required it.[7] If we go the second route, we might argue that the law does not always dictate how we should act. This is a question of policy, which is separate from law. Usually the right thing to do is to follow the law, but not always, especially in cases where law and morality conflict. Sometimes the morally right thing to do requires transgressing the current law because the law needs to be changed. This is clearly true in domestic law, and there is no reason to think it would be any different in international law. The appropriate analogy is the concept of civil disobedience. Before the Civil War, slavery was legal in the United States, but it was clearly wrong. Opposing slavery was the right thing to do, even though it was sanctioned by the law. Perhaps humanitarian intervention should be regarded in the same light as an example of civil disobedience. Although international law says that intervention to stop genocide is illegal, it's a case where civil disobedience of the law is morally justified.

We think that this view concedes too much. Admitting that humanitarian intervention to stop genocide is illegal under international law involves sacrificing the rhetorical high ground. Although moral arguments may be successful in the court of world opinion, one cannot walk into the Security Council conceding that a course of action violates international law and expect to prevail. The same is true in a domestic court; a judge will not be impressed if you admit that the law is not on your side but that you want him or her to rule your way regardless. It is true that you can sometimes get away with this with a jury by getting them to go with their gut and ignore the law altogether. But these cases of jury nullification always happen under the radar; the lawyer never says openly that he or she is requesting jury nullification. Juries have the right to nullify because no one knows what goes on behind closed doors. But lawyers do not have the right to ask juries to ignore the law. As officers of the court, they are bound by the law.[8]

Another strategy for defending humanitarian intervention is to look beyond the UN Charter to broader principles of international law, such as international humanitarian law, that might justify the use of force.[9] But the danger of these arguments is that they weaken the international prohibition on the use of force, which is the centerpiece of the international legal order.[10] Because the source of this prohibition is the UN Charter, any theory of humanitarian intervention must be focused on the Charter.[11] The argument should fall within the four corners of Article 51 and its standards for the use of force. Otherwise, one risks weakening the legal status of the Charter's prohibition on the use of force, which would have unfortunate consequences. The Charter would become less effective as a legal document that governs the use of military force in international law. And this weakening of the Charter would not be limited to cases of humanitarian intervention, where we might applaud the fact that the Charter is sidestepped in favor of more moralistic concerns. Once the Charter and the international system is weakened or delegitimized, it is weakened in all situations, including cases of less justifiable use of force, such as brute aggression. Once people start arguing that Article 51 is not the sole standard for justifying the use of force, the door is opened for all kinds of arguments about the use of force.

There are many ways to justify humanitarian intervention within the context of moral theory. One might make a simple utilitarian argument and claim that intervention is justified when it will save more lives than it will cost. If we can start a war that will cost only a few thousand lives but we will save millions of civilians from genocide, then the cost is certainly worth it. Although we have to violate another state's territorial integrity, this is a small price to pay for the great gains in utility maximization. The structure of this utilitarian argument would be the same for preemptive war. But regardless of the success of these consequentialist theories, they do not make much of a legal argument. Indeed, they are almost question-begging because they assume that maximizing utility is more important than respecting state sovereignty simply because nothing is more important for a utilitarian than maximizing utility. But this is precisely what we need to demonstrate if a theory of humanitarian intervention is going to be successful: that territorial integrity can be violated in certain extreme situations.

Some theorists offer a more explicitly legal argument for humanitarian intervention by appealing to the recent expansion in international humanitarian law. This field is supported by an explosion in recent scholarship on universal human rights. The structure of this human rights argument suggests that when a state government either purposively commits human rights violations or allows others to commit them and does nothing about it, then state sovereignty is forfeited. A similar move is made in domestic criminal law by theorists who argue that criminals voluntarily forfeit certain rights when they decide to act unlawfully. This explains why society has the right to inflict punishments and deprive felons of their liberty, conduct that otherwise would be impermissible. This human rights argument might even note that most states have signed international treaties obligating them to protect human rights, and even if they have not signed such a treaty, most of the human rights discussed in the Universal Declaration of Human Rights are considered to be nonderogable rights that function as *jus cogens* (peremptory norms). Individual nation-states cannot decide for themselves if they will respect these rights or not. They are universal rights. Although this avenue of inquiry appears promising, there is nothing in these human rights treaties that implies that states have abrogated their sovereignty if they fail to protect them. The basic principles of Westphalia still apply. And as for the human rights that are codified in the Universal Declaration, there was never any suggestion when it was adopted that it would be used as a justification for intervention.

Offering a convincing account of humanitarian intervention that appeals both to our moral sensibilities and to the existing legal framework is more than just a question of rhetorical strategy. It is a question of conceptual clarity as well. The goal of any theory of humanitarian intervention should be to adequately explain why and how such intervention is justified and in what circumstances, given that the use of force is so restricted by international law. There is no value in trying to commit an end run around the UN Charter and its Article 51 prohibition on the use of force. A successful theory will not only win the hearts and minds of the international community, but it will also convince international judges and members of the Security Council. It is for this reason that we must look to our theory of legitimate defense for a convincing rationale for humanitarian intervention. Our theory of legitimate defense works within the framework provided by Article 51 and links intervention with the general right of self-defense outlined in the Charter and recognized throughout the history of international law. The basic outline of the theory recognizes that the inherent right of self-defense belongs not just to member states of the UN but to national groups as well. When these national groups are attacked, they have the right under international law to defend themselves; moreover, other nations have the right to come to their defense consistent with legitimate defense. A systematic exploration of our account follows.

2.   THE RIGHTS OF NATIONS

There has been a long-standing tendency within international law to focus almost entirely on *states* to the exclusion of *nations*. This is not surprising, given that states are defined in very precise ways. They are geographically defined, constituted by governments (whether legitimate or not), and recognized by the world community as states with legal personality. In the context of international law, legal personality means that a state has the authority to engage in legal relations with other states. It can negotiate and sign treaties, it can borrow and lend money, it can appear before international bodies, and it can be held responsible for its actions. The state is a legal "person" in international law.

Nowhere is this more evident than in the International Court of Justice opinion regarding Israel's decision to build a defensive wall around the occupied territories.[12] After Israel started construction of the wall, the General Assembly asked the ICJ for an advisory opinion on the wall's legality. After concluding that the wall was inconsistent with international humanitarian law, the ICJ also rejected Israel's claim that the wall could be defended by virtue of Article 51 and Israel's inherent right of self-defense against terrorism. In language at once sweeping and swift, the ICJ rejected this argument, concluding that Article 51 had no relevance in cases where the armed attack comes from something other than another legal state, effectively limiting its application to attacks by a foreign state. If the attack cannot be attributed to a foreign state, according to the ICJ, Article 51 is irrelevant. Never mind for the moment that Article 51 simply refers to situations where "an armed attack occurs against a Member of the United Nations" and says nothing about where the attack must originate from. Under the interpretation proposed by the ICJ, Article 51 does not apply to cases of rogue terrorist activity unsupported by a foreign state. Of course, it may be the case that the wall was illegal for other reasons, although we leave this issue to the side. The ICJ's categorical rejection of Israel's self-defense argument is the most extreme example of international law's myopic fixation on states to the exclusion of other entities.

On the other hand, the *nation* is a confusing concept with inexact boundaries. We have a good idea of what people mean when they talk about nations, but just start to define one and you run into problems. What are the boundaries between nations? This is not clear. Nations have something to do with "peoples" and with "culture," this much is for sure, but how many nations do we have in our world? It is difficult to count.

But it is not difficult to count states. We can just look at the map or at the number of seats at the United Nations General Assembly. Of course, there are liminal cases when a state is recognized by some and not other members of the world community, or when a province considers itself an independent state but another country does not (such as Taiwan). Of course, not all states belong to the United Nations. Aside from these marginal cases, states are much more positivistic than nations. Nations are abstract, metaphysical entities that are difficult to ascertain. But states are easy to define because they have nothing to do with the way the world should be organized—they deal only with how the world is actually constituted, for better or worse. Whether or not Quebec should be its own state is irrelevant to figuring out how many states currently exist north of the United States. The answer is clear: one. No wonder, then, that international law prefers to deal with a clearly defined entity such as a state. States are members of the United Nations, can appear before the ICJ, have ambassadors, and have citizens that emigrate from one state to another. And states are responsible for their actions within the international community. These are all attributes of *states*.

Although international law has historically focused on states, it does not do so exclusively. International humanitarian law is playing an increasingly important role within the international legal order, as states are required to treat both individuals and groups consistently with human rights treaties, protocols, and declarations (both universal and regional) that they have signed. In many cases, these humanitarian standards protect national groups, including ethnic, linguistic, and religious minorities. It is becoming commonplace to talk of the cultural and political rights of minority groups, and these rights have legal status. Although these groups do not necessarily have legal personality within the system of Westphalian international law, their interests are nonetheless *protected* by international humanitarian and human rights law. In both peace and war, there are limits to how these minority groups can be treated, and failure to abide by these standards can result in extreme international consequences, including complaints before human rights commissions and cases before the ICJ. Similarly, the law of genocide is now a major part of both international law and international criminal law. It is an international crime to wipe out an ethnic group, and these prohibitions protect nations and national groups—not states. One might infer from these protections that nations have the basic right to exist under international law and that violating this right by killing individuals who belong to a particular ethnic group is the most serious breach of international law possible.

But nations and peoples have more than just a basic right to exist under international law. They explicitly have the right of self-determination as recognized by Article 1 of the UN Charter, which states that the purposes of the United Nations include promoting "friendly relations among nations based on respect for the principle of equal rights and self-determination of peoples." This right belongs to all nations and peoples.[13] The International Covenant on Civil and Political Rights is even more explicit. It states, "All peoples have the right of self-determination. By virtue of that right they freely determine their political status and freely pursue their economic, social and cultural development." This right also includes the right of all peoples to "freely dispose of their natural wealth and resources," and a people cannot be "deprived of its own means of subsistence." The right of self-determination also has a long history in customary international law.[14] Finally, state parties of the ICCPR are obligated to promote "the realization of the right of self-determination." In addition, Article 55 of the UN Charter specifically refers to self-determination, and Chapters 11 and 12 on decolonization certainly imply the concept as well. Indeed, these principles of self-determination were critical to the basic deal that led to the creation of the post–World War II UN system. Because the UN system placed such a premium—some would call it a fetish—on world stability, the world's minor powers worried that the system would simply entrench colonial and regional dominance by the world's major powers. Because territorial integrity was to be protected at all costs and any deviation from this principle was viewed as illegal aggression, the status quo would be maintained and protected by the international legal order, even if that meant that fundamentally unjust arrangements persisted. It is for this reason that the principle of self-determination was explicitly recognized in the Charter. It was especially important for nations who had recently gained independence from their colonial masters, and was meant to codify the right of peoples to rebel against colonial domination and fight for self-rule.

This right of self-determination could not have been attributed to states. That is a logical impossibility. To understand this point, we must first understand what this right means. In a sense, the right of self-determination is about autonomy—autonomy to determine one's own political, economic, and cultural affairs. Usually, the most appropriate avenue for this autonomy is a state, through which one can establish secure boundaries, run schools and other cultural institutions, and legislate and enforce laws in the public sphere.

If only states had the right to become states, the right would devolve into a tautology. But peoples and nations that are not yet states have the right to determine their future and organize themselves as a state to make it happen. This formulation of the right has some meaningful content. International lawyers are somewhat insensitive to these distinctions, if only because matters of self-determination are rarely decided by international courts or tribunals. One does not go to court to determine if a nation should be recognized as a sovereign state by the international community. The international community decides these matters on the basis of more practical considerations: if a nation has de facto control over its territory, it is considered a state; if not, then not.

The right of self-determination is therefore the right of a people—however one chooses to define it—to be free from domination and have the necessary authority to control their own affairs.[15] Usually, although not always, this means being in control of their own state, which allows them to govern themselves and organize a government around their core principles, assuming of course that they respect universal human rights in the process. In the case of a smaller ethnic group living within a democratic state, self-determination may mean that the political structure is such that they have enough regional autonomy to organize their social and political lives.[16] At the very least they must have access to the political system so that they are adequately represented in the state's government and can use this electoral influence to press their claims. This is a right that attaches to *peoples*.[17] It is in this sense that one might say that states do not have a right to self-determination. Peoples or nations have the right, and it frequently involves the creation of their own state. A state is usually, though not always, the *fulfillment* of that right. Or, more precisely, the state is the *vehicle* through which the right of self-determination is usually exercised.

Nationalism is the pursuit of self-determination by a nation.[18] Nations advocate for their political self-determination by arguing for—or fighting for—their own state. These are nationalistic endeavors, and they are frequently supported by nationalist ideology and nationalist rhetoric. In the most general sense, then, nationalism is the idea that there should be a one-to-one correspondence between nations and states, that a nation must be in control of its own state in order to exercise its own autonomy.[19] Sometimes nationalistic movements are completed in the political sphere. When they are not, they often transform into terrorism, guerrilla war, rebellion, or revolution, waged against the prevailing governmental powers from whom independence is sought.

The question of secession is a difficult one. When is a national group justified in waging a war of independence against a government that refuses to recognize its political authority, or refuses to allow a democratic referendum to decide the question of separation? It is clear that national groups have a right to meaningful participation in the political process, but it is not clear how far this right extends. On the other hand, it is equally clear that national groups do not always have the right to secede; this would create a world of chaos.[20] There is an important distinction here between the morality of secession and the legality of secession. Allen Buchanan, among others, has written extensively about the moral right of secession.[21] He argues that a moral right to secede exists for minority groups, but only when certain conditions are fulfilled. Buchanan situates his argument within the sphere of political theory, and his work has spawned much attention among political scientists. Strangely, though, the right to secede is underdeveloped in international law, which has little to say about the legality of secession, except that it is in principle justified by the principle of self-determination. All that matters is whether other countries recognize a new nation after it attempts to secede. For example, during the Civil War, the North refused to recognize the South's attempt to leave the Union. But when are these separations justified, and when are they illegitimate? And who gets to decide?

Locke argued that the social contract between citizen and government is based on a fiduciary duty.[22] The government is required to lead so that the citizens can have their basic rights (such as life and property) protected and to guarantee those natural rights that could not be secured in a state of nature. But when a government betrays that fiduciary relationship, the contract is breached and the citizens have the right to cast off their chains and launch a rebellion. This decision can be made only by the citizens; who judges whether this rebellion is justified or not? If any citizen can claim a breach and rebel against the government, it would seem that no government could enforce the loyalty of its subjects. So there has to be some kind of limit on the power of rebellion. Locke had a very instructive answer to this question. He argued that when a government fails to protect the basic rights—life and property—of its citizens, they are justified in waging war against their own leaders. This decision can be submitted to two judges. The first is the world community, to whom we must justify our actions. A national group that rebels against its government must submit its actions before the court of world opinion for adjudication. There is a certain logic to this idea. If a national group wins a war and declares its independence, the world community must decide whether to recognize the group as a state with legal personality, or whether they will consider it a rogue province that has no independent legal authority. So there is a very practical reason why these disputes must be submitted to the court of world opinion. Only the community of world states has the power to admit new members to the community of states that can negotiate and sign treaties and engage in all of the international interactions that come with statehood.

The U.S. Declaration of Independence is a perfect example of this phenomenon. The document explicitly outlines a list of grievances against the king, from taxation to the lack of jury trials, to justify the rebellion and urge the world to recognize the new United States. The list of grievances in the Declaration was not meant as a petition to Britain for redress. The United States had already declared its intention to separate, and the time for redress had passed. The Declaration was meant to justify our reasons for rebellion to the world community.

According to Locke, the second judge of a rebellion is God. Citizens who engage in a treasonous rebellion will receive their just deserts from their maker during the final judgment.

Debates about nationalism often focus on the difficulties of defining the nation. Clearly, the following elements are important: race, religion, ethnicity, language, a common culture, and geography. But these factors often intersect with each other, making the boundaries between nations unclear. Do the Basque belong to the same nation as the rest of Spain? If not, why not? Is it because they have a different language or a different history? What role does Roman Catholicism play in the analysis? Although these problems of definition are perplexing and have spawned an entire academic literature devoted to the subject, we should not let this obscure the basic fact that we have a commonsense understanding of what it means to be a nation, even though it is difficult to define.

It is quite possible that the idea of a nation is a cluster concept, composed of more primitive ideas such as race, ethnicity, language, and culture. Although this may be true, it does not necessarily mean that the idea of nationhood is incoherent or must be rejected. Just because the internal structure of "nationhood" is confusing and may not be atomic is no reason to entirely jettison the idea.[23] We ought to be pragmatic about our choice of concepts. And the concept of the nation accords with our commonsense understanding that national groups have a right to self-determination and to be free from colonial rule.

It is difficult to identify nations based on hard criteria or necessary and sufficient conditions. For example, we think of the Swiss as a nation, even though they do not share a common language and were historically a loose confederation more than anything else. Also, the Jews were clearly a people before the creation of Israel, with a common language, religion, and history, even though they lacked geographic continuity and were dispersed worldwide. This, of course, does not mean that either the Swiss or the Jews are not a nation. Rather, this proves the reverse: we cannot always pick out nations or peoples based on necessary and sufficient conditions.[24]

It may be more accurate to say that nationhood is a family resemblance concept. Wittgenstein argued in the *Philosophical Investigations* that the concept of a game is a family resemblance concept.[25] We all know about games: the word encompasses everything from checkers and soccer to crossword puzzles and solitaire. But it turns out that constructing a definition of games is actually quite difficult. Most definitions fall victim to being either under- or overinclusive. But none of this suggests that we do not understand what people are talking about when they refer to games. Rather, it suggests that we define games based on core examples that we know are games (checkers and chess) and then start including other examples that have features or qualities similar to those of checkers and chess. We group individuals together into families based on these similarities. These groupings happen not by iron-clad criteria that we can enunciate in a general and abstract way, but by inductive reasoning and analogy. Nationhood fits into this category of family resemblance concepts. We have a definite idea that Russia and France are nations, and we work our way out from there. Other groups share features similar to those of Russia and France—common languages, a common sense of identity—and so we call them nations, too. And so on and so forth. It is therefore imperative that we take seriously this concept of nationhood, even if international law would rather replace it entirely with the concept of statehood.

Another place where nationhood matters is our collective moral psychology. Individuals believe their identity is constituted, at least in part, by their participation in a collective national group. In cases where this national group coincides with a nation-state, the psychological identification may target the state or the government. But this is not always the case. Individuals often feel strongly connected to each other through their common participation in a national group, even when the national group has no control over a nation-state or cuts across state boundaries in some uncomfortable way. Basque individuals think of themselves *as* Basque, and this perception forms a significant part of their sense of self, their values, and their identity. They value their participation in the Basque way of life, and if they are blocked from this way of life, they feel wronged.

The moral psychology of nationhood also manifests itself in feelings of pride and resentment.[26] As individuals embedded in a national culture, we feel pride in the accomplishments of our nation. This is not just a function of soccer matches and the Olympics; it is also a function of pride in artistic achievements, a noble history, or literary greatness. We also resent national groups that have engaged in less noble conquests. Consider a Jewish survivor of the Holocaust or a Tutsi survivor of the Rwandan genocide. They may not only resent individuals who are, say, Germans or Hutus, but they may also resent Germans or Hutus *as a group*—as a nation. Now one must be careful here when thinking about these moral reactive attitudes. A member of an oppressed national group may decide, as a matter of reason and logic, that he is wrong to think such thoughts about a nation as a whole, and that only the individuals who committed these wrongs (Nazi officers, etc.) should be held responsible for their actions. But these rational feelings come into play only after our moral reactive attitudes have emerged. The point is that we cannot help but feel pride in our nation's accomplishments and resentment at another nation's harmful conduct, even if we are good Enlightenment liberals who believe that only individuals matter. Our moral reactive attitudes are open to only so much revision.

This territory of collective pride and collective guilt is extremely perplexing. One might be tempted to argue that it is a fiction, a mere illusion that should be discouraged and replaced with a cosmopolitan commitment to democratic institutions, not Romantic notions of the nation. But one must take seriously the moral psychology of nationhood, especially where collective shame is concerned. The crimes of one's nation can result in a kind of collective shame that extends downward to individuals who never participated in the crime, or were not even alive when it occurred. It is in this way that young Germans can feel ashamed of the Holocaust, because it was a crime committed not just by a few Germans, but by Germany itself. Similarly, some Americans feel ashamed by slavery or by the systematic elimination of the Native Americans, even though these crimes were committed long ago. Crimes committed by nations linger in the collective psychology of those who belong to the nation. We explore the source of these feelings in chapter 8.

These national ties may be illusionary in some respects, but they are significant nonetheless.[27] Nations often exaggerate their homogeneity by rewriting their history, emphasizing unifying elements and deemphasizing centrifugal forces. The notion of a common and unidirectional culture is, to be sure, a fiction. In reality, cultures are fragmented and discordant, full of opposing forces and voices that cannot all be assimilated.

Many of our ideas about the nation come from an analogy to the individual, or the person. We think of nations on the world stage as individuals writ large, interacting with other such individuals, to which we can attribute mental states, intentions, beliefs, and desires, in much the same way that we attribute beliefs and desires to individual persons. Underlying all of this is an assumption that nations are rationally unified in the same way that individual persons are rationally unified. This analogy goes back at least as far as Plato's *Republic*, if not further.[28] However, this is problematic as an analogy, and it may not even be totally true about individual persons. Individuals are less rationally unified than we would like to believe;[29] we are each full of discordant voices that make up our fragmented self. This is part of the illusion of the classic Enlightenment view of the rational self.[30] It is therefore probably true that we exaggerate the rational and psychological unity of both nations and individuals.

Nevertheless, these "imagined communities" play a significant role in our lives. Whether based on hardened fact or illusion, individuals situate themselves in national cultures that form a major part of their sense of self and their psychological reality. It is therefore crucial to recognize the importance that nations play, not only in our individual lives, but in our interactions with the world. It is often nations that go to war with each other, engage in ethnic cleansing, and defend themselves against genocide. Although international law may be primarily concerned with the actions of nation-states, it is nations that are the more primary unit by which states are eventually composed. And any theory of international self-defense must be sensitive to this fact.

3.   DEFENDING THE NATION

We must now return to our core dialogue about the right of self-defense. Having discussed the distinction between nations and states, we must ask: To whom do we attribute an international right of self-defense? Does it belong to states or to nations? The general consensus among international lawyers is that the right belongs to states. First, Article 51 refers to member states of the United Nations. Second, post-Westphalia international law has always concerned itself with the right of states, to the exclusion of other entities. The issue is considered so obvious among international legal scholars that it is never addressed. So our task here is to raise a few critical points showing that the obvious is actually not so obvious. There is reason to believe that a deeper understanding of international law will recognize a right of self-defense belonging to *nations*.

First, Article 51 of the Charter describes the right of self-defense as inherent, or in French, *le droit naturel*. This means of course that the right is not created by the UN Charter, which simply recognizes it. It existed before the Charter came into force, and it was not created by treaty obligations. It is inherent. If this is indeed the case, then we must think seriously about this underlying inherent right and where it comes from. Although the English language of the Charter gives us little guidance about the nature of this right, the French language is instructive. The right is a function of natural law, in the highest sense of the expression, not man-made law.

To whom does natural law grant the right of self-defense? Again, an analogy to the right of individual self-defense will guide our analysis. The right of individual self-defense exists in natural law because it is the right that all individuals enjoy in a state of nature before a social contract creates a functioning government. Similarly, one might think of nations as having existed in a state of nature with each other, forming a social contract that created the Westphalian international legal order. But nations retain the inherent right to self-defense, just as individuals do.

This analogy between nations and individuals forms a crucial part of our understanding of the international legal order. John Rawls argued in *The Law of Peoples* that peoples bargain together in a second-tier social contract to create the basic institutions of international relations, including the institution of statehood and the scope of sovereignty.[31] Peoples—or what one might call nations—form the basic unit for the second-tier social contract because nations, just like individuals, are moral agents, capable of forming agreements and being responsible for their actions. According to Rawls, this hypothetical agreement among nations, in the most abstract sense of the expression, justifies the existence of the modern system of nation-states.

Rawls develops an interesting definition of "peoples." He argues that they have three features: a just constitutional government, a unity based on "common sympathies," and a moral nature.[32] "Common sympathies" is a phrase borrowed from John Stuart Mill, meant to evoke the variety of attachments—language, religion, ethnicity—that bind a culture together. For the moment, it does not matter what these elements are. For each people the common sympathies might be different. Also, it does not matter if these common sympathies stem from illusion or myth. All that matters is that the common sympathies "make them cooperate with each other more willingly than with other people, desire to live under the same government, and decide that it should be government by themselves, or a portion of themselves, exclusively."[33]

Once we link this notion of nations bargaining together to agree on the basic principles of the international order with our previous point that the right of self-defense is *naturel*, then it becomes clear that the right to self-defense extends not just to the modern system of the UN Charter, but indeed extends to the very foundations of the international social contract. In an international state of nature, all nations have the right to defend themselves, and this is the source of the right recognized by Article 51 of the UN Charter.

This last step provides the general outline to our argument. Because nations have a natural right to self-defense in the face of an armed attack, it follows that others have a right to come to their defense as well. This should not be a shocking conclusion. Our previous account of legitimate defense involved a unified approach to self-defense and defense of others under a single umbrella, with a common set of conditions for the exercise of both. It is a natural conclusion of this account that any individual has the right to come to the aid of a victim who would be justified in using self-defense with his own hand. Similarly, the world community can come to the aid of any nation who has a legitimate claim to self-defense against an armed attack.

This forms the basic structure of our defense of humanitarian intervention. When a nation is not in control of its own affairs but is part of a larger state and it suffers an attack against its own interests, either from its own government or from outside forces, it has an inherent right of self-defense. And the world community has the right to intervene on its behalf through the exercise of legitimate defense.[34] Moreover, this exercise of legitimate defense falls within the parameters of Article 51 of the Charter. This is important, because most attempts to justify humanitarian intervention sidestep the Charter by appealing to exterior legal norms. For example, the report of the International Commission on Intervention and State Sovereignty appealed to a principle called "the responsibility to protect" that, when violated, might justify foreign intervention.[35] Although the principle is somewhat grounded in international humanitarian law, the report's analysis of its relationship to the UN Charter's Article 2 prohibition on the use of force is decidedly weak. The report simply notes that if the Security Council fails to act, the General Assembly has secondary authority to secure peace and security.[36] However, this authority under Article 11 of the Charter is limited to nonbinding recommendations. This, or any other legal norm, is hardly sufficient to overcome the explicit directive in the Charter—the highest expression of binding law in the international system—that Security Council authorization and the inherent right to self-defense are the only two circumstances that justify military action.

In contrast, our defense of humanitarian intervention works within the four corners of Article 51 and explains how intervention can be justified in an international legal order obsessed with state sovereignty. By appealing to Article 51's language of droit naturel and légitime défense, we can understand that all states have the right to exercise military force in defense of a nation whose existence is threatened by an armed attack, even if this intervention requires the violation of another's state sovereignty or territorial integrity. Indeed, it is always the case that using defensive force to protect a victim will involve a physical transgression against the aggressor. This is precisely the point of legitimate defense.

4. PROBLEMS

Our defense of humanitarian intervention, grounded in our theory of legitimate defense, is not without its problems. We must now address its shortcomings and demonstrate that, despite them, legitimate defense offers the most compelling account of foreign intervention under Article 51.

The first obvious point is that our account of humanitarian intervention, like all other accounts, flies in the face of the plain language of Article 51, which limits the right of self-defense to member states of the United Nations. We have gone to great lengths to demonstrate that the right belongs also to nations, not just member states of the UN. However, one might reasonably respond that the language of Article 51 is clear: the signatories of the UN Charter granted the right to member states only, and had they wanted to codify a broader right they would have certainly done so with more explicit language. But they did not, and the language of Article 51 is clear that the right belongs only to member states.

This objection shows an insufficient attention to the structure of Article 51. The article specifically states, "Nothing in the present Charter shall impair the inherent right of individual or collective self-defence if an armed attack occurs against a Member of the United Nations." Article 51 is therefore simply a limitation on the Charter prohibition against the use of military force. It says little about the full scope of international self-defense. It simply reaffirms that member states need not wait for Security Council authorization before they defend themselves against an armed attack. What rights are held by non-member states—whether natural rights or a positive right under international law—is unclear. It is therefore an exaggeration to interpret Articles 2(4) and 51 as forming a blanket restriction on the use of military force. It would be more accurate to see Article 51 as a limited entitlement granted to member states, while leaving the rest unstated. When combined with the explicitly recognized right of self-determination, the language of Article 51 is in fact very compatible with our account of legitimate defense.

The second possible objection to our view is that it relies too heavily on antiquated notions of race and ethnicity in defining nations. The use of racial and other ethnic criteria to define nations offends our Enlightenment sensibilities. We have moved beyond Romantic notions of blood ties and tribal nations and replaced them with cosmopolitan political communities. Neither race nor religion is destiny, and we can all choose the political ties that bind us together. To suggest otherwise is to take race and ethnicity—immutable characteristics both—and bind the choices of free rational agents.

This objection has a kernel of truth, but it exaggerates the racial quality of our notion of nationhood. First, nations are defined based on common connections such as language and religion and a common cultural heritage. There is no reason that these connections need be racially defined. While it is certainly true that members of a nation with a common cultural heritage and a common religion may also share a common race, it is not necessarily so. Second, the objection exaggerates the degree to which these national ties can be simply disregarded and replaced with Enlightenment political commitments. But this totally ignores the fact that many individuals consider their participation in a common national culture (whether built around language, religion, or history) to be essential to their personal flourishing. The meaningfulness of this participation in a common national culture cannot be swept away by fiat.

This leads to another objection about any account of humanitarian intervention that is built around nations. National groups are rarely unified across all categories: language, religion, race, and history. One can always divide up a national group into ever smaller ethnic groups by appealing to other categories. The Italians speak Italian and are Roman Catholic, but regional differences remain, each with a separate historical heritage. Does each qualify as an independent nation? The French are united by their language, but if we insist on unification by religion as well, does this mean that we should divide up the French along religious lines?[37] Do we need to divide up the British between Protestants and Catholics? This would seem to suggest an infinite regress with increasingly small ethnic groups with pure homogeneity and the same skin color, religion, and language. If we follow this path we end up with many ethnic collectives smaller than the nations we were originally hoping to define.

These problems of definition strike some as insoluble and a reason to abandon a Romantic account of nationhood in favor of a constitutional one. In countries such as Germany, Canada, and the United States, the unifying elements of the national culture are a commitment to basic principles of constitutional identity. In the United States, one might say that a belief in the U.S. Constitution, and in particular the protections of the First Amendment, form the organizing principles around which our constitutional nation is defined. It doesn't matter where you were born and which God you worship—what matters to be American is that you pledge allegiance to the Constitution and the legal protections of the Bill of Rights, including freedom of speech, religion, assembly, thought, and conscience. These are the core principles that we all share as Americans. One might tell a similar story about the Canadians, who share a commitment to the social welfare state, including nationalized health care, and a tolerance of minority cultures. Germans are loyal to their constitution and the post–World War II political system that they created as a rejection of their Nazi-era racialism. In all three of these examples, skin color, language, and race take a back seat to political freedoms as the organizing ideas of the nation.

Nothing in our account of nationhood requires that all nations be defined without regard to bedrock constitutional principles. In the number of nations that have successfully created a cosmopolitan nation—particularly those with a strong commitment to immigration and religious and linguistic freedom—it is correct to note that there is no ethnic unity to these nations. It may some day be the case that all nations will be organized around these key liberal commitments. But it is an exaggeration to suggest that all nations follow the American model here. Most nations are still unified by a common cultural heritage with an ethnic component, and even the great constitutional democracies of Europe—France, Germany, Italy, and others—are built around a common ethnic heritage that plays a strong role in their national identity in addition to their constitutional and political commitments.

There is a third objection. Occasionally the racial and ethnic criteria of a national group are not defined by its own members, but imposed by outside forces. For example, the distinction between the Hutu and the Tutsi in Rwanda was largely insignificant until Belgian colonial rule in the nineteenth century. Although there were Hutus and Tutsis before then, the distinction was not culturally relevant, and the two groups were integrated. But when the Belgians showed up with their nineteenth-century pseudo-science, they took out their rulers and started measuring foreheads and noses and making grandiose generalizations about the two ethnic groups. They claimed that Hutus were physically stronger, while the Tutsi were more intellectual and suited for white-collar tasks. The Belgians organized Rwandan society around these roles, which persisted even after Belgian colonial rule over Rwanda ended with the country's independence in 1962.[38] This means that the Tutsi nation is, in many ways, a category that is not internally self-generating, but rather imposed externally by outside forces.[39] Furthermore, this imposition of group status was arbitrary and based on fiction and prejudice.

The arbitrariness of these definitions need not trouble us. Even if national identity is externally imposed by an outside group motivated by a desire to oppress another group, the oppressed group—arbitrary though it might be—is nonetheless held together by how it is viewed by the rest of the world. Their common persecution may be the most pressing commonality they share. More specifically, the national group negatively defined may have, ironically, the best case for nationalism, because they must exercise self-determination to create their own nation-state that will protect them from groups that seek their destruction. Regardless of the origins of the Tutsi national group, this group status was the reason they were targeted and that a genocide was launched against them.[40] For this reason they have the right to defend themselves.

The Genocide Convention restricts the definition of genocide to conflicts among "national, ethnical, racial, or religious" groups.[41] At first glance, it is unclear if the Tutsis meet this definition. It is difficult to precisely identify the exact nature of the distinction between Tutsis and Hutus. The best candidate is the broad category of ethnic or "ethnical" differences, in the words of the Genocide Convention, although it is hard to define ethnicity in such a way that it covers the social differences imposed on these two groups by the Belgians. But these difficulties need not concern us, for we ought to look to the spirit of the Genocide Convention, its birth from international disgust following the Holocaust, and the international community's concern with the killing of a human "genos."[42]

We can see now the role that genocide plays in the argument about humanitarian intervention. Ethnic groups have the right to be free from oppression and genocide. They have the right to *existence* in the most basic sense of the word and the right to defend themselves in the face of any attempt to use military force to destroy them, regardless of whether or not these attacks cross an international border. These rights are droits naturels. Self-defense is therefore the primary notion in our analysis. Finally, the rest of the world has the right to come to their aid as a function of legitimate defense. And this right is entirely consistent with Article 51 of the UN Charter. Humanitarian intervention is therefore justified as a limited case of legitimate defense.

It is also clear now why state consent is not necessary in a case of defense of others.[43] In cases of humanitarian intervention, the nation being defended may be the victim of genocide by the very state whose borders are being crossed. It would therefore be absurd to require state consent before defense of others is undertaken. One might argue that the nation that is defended—say, the Kurds or the Kosovars—should issue the consent. But the victimized national group may have no official channels through which it could offer such consent, neither an assembly for democratic decisions nor a diplomatic channel to communicate those decisions. Remember, after all, we are dealing here with a *nation*, not a state.

5. APPLICATIONS

Having outlined our account of humanitarian intervention, we should return to the various examples discussed at the beginning of this chapter.

The NATO bombing that forced Milosevic and the Serbs from Kosovo was never authorized by the Security Council. As noted, this leaves two possibilities: either the bombing was justified by self-defense or it was illegal under international law. Our doctrine of legitimate defense confirms our intuition that the intervention in support of Kosovo province was justified. Although Kosovo was not recognized as a state with international legal personality, it was a defined geopolitical unit, represented by a coherent ethnic group seeking self-determination. Moreover, Kosovo's population was under military attack because of its ethnic identity. If the *inherent* right of self-defense is to have anything other than a purely formalistic meaning, it must apply to an ethnic group seeking protection from ethnic cleansing and genocide. The fact that the Kosovars did not yet have their own state should have no bearing on our analysis. All that matters is that the Kosovo Albanians were a coherent enough ethnic group that another ethnic group sought their destruction. The Kosovars, as a *people*, had a natural right to defend themselves against genocide when the Serbian-supported militia started going door-to-door to collect ethnic Albanians for displacement, rape, and extermination. The doctrine of legitimate defense suggests that not only did Kosovo have the right to defend itself by launching a war of self-defense, but the world community had the right to intervene on its behalf as a legitimate exercise of defense of others to prevent the Serbs from annihilating the Kosovars. And all of this was consistent with le droit naturel de légitime défense.

Similar arguments apply to Rwanda. Had Clinton or other world leaders been motivated to act, our doctrine of legitimate defense would have justified an intervention on behalf of the Tutsi. Facing extinction at the hands of Hutu Power militia, the Tutsi nation had a natural right of defense against genocide, including the right of self-determination to create their own nation-state where their right to existence could be protected. By linking the inherent right of self-defense (in the UN Charter) with the right to be free from genocide (in the Genocide Convention) and the right to self-determination (in the Universal Declaration and the ICCPR), it is clear that the rights of Tutsi were internationally recognized. Clinton could have launched a military attack to disarm the Hutus and engage in a legitimate defense of the Tutsi. Unfortunately, this was not to be the case, and the genocide continued without delay.

The case of Somalia is somewhat more complicated. Mogadishu was not gripped by warring ethnic groups, but rather rival warlords who ruled as criminal thugs with private militia. Loyalty was assured based on pragmatic considerations, such as food and money, rather than ethnic heritage. Consequently, it would seem that our doctrine of legitimate defense would not justify intervention on behalf of one warlord's militia over another. Human rights advocates justified the intervention in Somalia on the grounds that civilians had no access to relief supplies, including food and medicine, because these were being blocked or stolen by the militias. But it is unclear if we can point to a nation or people in Somalia exercising legitimate defense against an armed attack either from within or without. The only possibility would be the Somali people themselves, although this seems implausible as a matter of self-defense. The situation seems best described as a failed government that collapsed into anarchy and civil war. It is therefore possible that our notion of legitimate defense does not justify humanitarian intervention in this situation. Some might count this as a defect of our account because it applies in only narrow factual circumstances. But far from being a defect, this only demonstrates that our account of humanitarian intervention is based on a theory of self-defense that is cognizable under international law, and the facts in any given circumstances must be scrutinized to determine if they fit the parameters of our account. The doctrine of legitimate defense is indeed a narrow one, but it is better to provide an account of humanitarian intervention that is legally convincing than a broad one that applies in many factual circumstances but has no basis in international law.

Finally, consider the case of Iraq. Although the stated rationale of the Bush administration was the existence of weapons of mass destruction, an alternative rationale, stated by some intellectuals,[44] was the protection of Kurds and other ethnic minorities from Saddam Hussein's wrath. Hussein gassed the Kurds in 1988 and was eventually put on trial for genocide, though he was executed for crimes against humanity before a verdict could be reached in his genocide trial. Nonetheless, the Bush administration might have argued that the invasion was justified to protect the Kurds and other minorities (such as the less favored Shiites, who were the frequent targets of Hussein's policies) and that an invasion was an exercise of legitimate defense. Unfortunately, there are two problems with the application of the doctrine in this circumstance. First, it is clear that the U.S. government was not motivated by a desire to protect the Kurds. This was an ex post justification that was championed by liberal supporters of the invasion, but it had little bearing on the Bush administration's conduct, which seemed more concerned with the possibility that Hussein might use weapons of mass destruction against the United States. Second, there is no evidence that an attack by government forces was imminent. Hussein had launched a genocidal campaign against the Kurds in 1988 and also put down a Kurdish uprising after the first Gulf War. The Americans allowed Hussein to keep his helicopter gunships, which he used to suppress Kurdish forces in the north. At that point, the United States might have been justified in invoking legitimate defense to join the Kurds in their war of self-defense against Hussein. But the United States at the time had no interest in toppling Hussein and no appetite for more military action in Iraq. The Kurdish uprising was suppressed. One cannot then retroactively invoke legitimate defense a decade after the original attack.

7

PREEMPTIVE AND PREVENTIVE WARS

When criminal lawyers talk about preemptive self-defense, they mean something negative, but when international lawyers use the same term, they mean no more than that the defender used force before it was faced with an "armed attack." Although the doctrine of preemption gained renewed notoriety following the U.S. invasion of Iraq, the issue is one of the oldest in international law and has been debated ever since legal scholars started discussing the legality of warfare.[1] The early debates focused on the balance of power between states and whether a nation could prosecute a war merely to restore its regional superiority or prevent a neighboring nation from increasing its military strength. But in the modern framework of the UN Charter, it is the phrase "armed attack occurs" that draws the line between legitimate and illegitimate military force. A military response in the absence of an armed attack is illegal under international law unless authorized by the United Nations Security Council. But no one could reasonably propose a doctrine of self-defense that was limited to striking back only after being struck by a phalanx of bombers or guided nuclear missiles. Indeed, there would be an obvious contradiction in requiring states to wait until the missiles were under way or had struck their targets. Justifiable self-defense would then automatically qualify as a reprisal, inflicted after the attack was over. This would turn the entire notion of self-defense on its head.

1. REDEFINING IMMINENCE

Preventive war would normally be disallowed by the doctrine of legitimate defense, because an attack must be *imminent* before the right to legitimate defense is triggered. This was discussed in chapter 4. Unfortunately, international law gives us little guidance on the question because there is no treaty explicitly defining imminence. And although it is clear that the imminence requirement is part of customary international law, we have no idea what the term means when it is used by state parties. Therefore, the battleground for this debate is the correct definition of imminence. Do the soldiers have to be crossing the border for the attack to be considered imminent? Surely this imposes too high a burden on the defender. Under these conditions, the defender would never have the opportunity to muster sufficient resources to successfully respond to the attack. On the other hand, if we define mere preparations for war as within the definition of an imminent attack, we grant the defender license to attack every strategic enemy.

Kant, surprisingly, argued in favor of preemptive war, suggesting that any preparation for war represented a shift in the balance of power that constituted an act of aggression. Kant argued that the international system is essentially a state of nature, with no overarching legal system that could vindicate a state's rights. "States, like lawless savages, exist in a condition devoid of right," he wrote.[2] One state cannot sue another state for violating its interest, and so it falls on each state to use force when necessary to protect its interests, just as each individual can protect himself in the state of nature. Furthermore, mere preparations for war might constitute an act of aggression, according to Kant, against which another state might legitimately claim a "right of prevention" that it might vindicate with its own forces. Similarly, any attempt by one state to alter, to its advantage, the balance of power in its region would trigger the same right of prevention. Kant says that there is a "right to a balance of power among all states that are contiguous and could act on one another."[3]

For a philosopher concerned about "perpetual peace"—the end-game that we are all striving for—Kant's view strikes most modern readers as a recipe for constant war.[4] However, his views are instructive and demand greater attention. Kant thought that our ultimate goal should be to achieve a perpetual peace, though he did not think such a state of affairs was possible in the international realm. The only way to achieve perpetual peace would be to create a world government through a social contract among the world's nations. Though an enticing possibility, the globe is too fractured with diverse and warring parties to be unified under a single government. The most that can be achieved, according to Kant, is something *approaching* world government through a weak confederation of states, committed to peaceful relations with each other and providing a legal framework for resolving disputes. These arrangements are entirely voluntary and can be dissolved at any moment, throwing states back into a state of war with each other.

Here is the interesting part of the argument. In the state of war, a nation has the right to vindicate its rights, "unlimited in quantity or degree," against an unjust enemy, since there is no world government to do this for it.[5] This is literally "the right to make war," and it includes launching a preemptive strike against military preparations. How does one decide who is the unjust enemy? Kant's answer is highly instructive. Borrowing on the structure of his great categorical imperative, Kant argues that an unjust enemy is one who expresses a maxim that "would make peace among nations impossible and would lead to a perpetual state of nature if it were made into a general rule."[6] This mirrors the same kind of universalization that he urges us to adopt when we decide if our own individual actions violate the categorical imperative.[7] Ironically, this is the very problem of preemptive war. If every state prosecuted its strategic interests by launching preemptive attacks, the world would indeed collapse back into a state of nature. This is the very reason for the UN Charter's general prohibition against the use of force.

Of course, much has changed since Kant's time. There is now a functioning international legal system, including the United Nations, which did not exist when Kant was writing. It can no longer be said that states are in a state of nature with each other. Indeed, states *can* sue each other, not just at the International Court of Justice, but also at other regional and international organizations devoted to particular subject matter, such as the World Trade Organization. True, it is important not to exaggerate the scope of the international legal system. Much of international law must be enforced by states themselves, and there is no international police force to execute judgments. But this is a facile point. Questions of international law are capable of adjudication, at the United Nations and elsewhere, and states have access to numerous legal fora to resolve these disputes.

The real issue here is imminence. A state is not required to wait for adjudication of its disputes when it is looking down the barrel of a gun. Whatever vindication that might be had would come much too late. But this is also true in the domestic law of self-defense. Indeed, this is the whole point of self-defense. Imminency requires that you stand in the position of the government and use the force that they would if they were on the scene. But you cannot wait for a court to recognize your rights when the danger is imminent. No court can give you your life back once it is taken from you. It is therefore essential that you act immediately. In a sense, then, you temporarily borrow back from the government the authority to exercise force that you ceded to the government through the social contract. The same is true of the state, which is prohibited from using force under the UN system, but may do so when faced with an imminent attack.

The question is how to determine when an armed attack has begun. Any coherent doctrine of self-defense would permit a response as soon as the attack becomes imminent. Sensible interpretations of the Charter read the word "occur" to mean "be actually present," and "present" means that the attack is imminent or about to occur. Thus, in his authoritative study of self-defense under international law, Yoram Dinstein comes to the conclusion that the Charter permits defensive force against an imminent attack.[8]

Whether you call a military response preemptive or not depends on your definition of a military "attack." There is nothing preemptive about responding to a military attack in progress, but how we should define an attack in progress is totally unclear. So in terms of justifying real-world military responses, the concepts of attack and preemptive war are inversely proportional. If you have a sufficiently broad definition of attack, then you may not even need to resort to a theory of preemptive war, because you can simply claim that the military response is designed to stop the attack, not prevent it. If, on the other hand, you have a narrow definition of what constitutes an attack, then a good theory of preemptive war is essential, because you will need to find a justification for launching a military response before the attack began. Many commentators are confused on this point. You do not need a broad notion of what constitutes an attack and a theory of preemptive war; one or the other will do.

At first glance it might seem easy to come up with a workable definition of what constitutes an attack. It goes without saying that when the troops are crossing the border and the planes are in the air, we have an armed attack. But what if the military forces are mobilizing but have not reached the border? What if the military forces are just training for the attack? What if spy planes are gathering intelligence that will be necessary to launch a successful attack? The question is where to draw the line. Once you admit that a state need not wait until the troops are crossing the border before engaging in legitimate defense, the question of line drawing becomes very difficult.

Perhaps the most famous example of preemptive engagement was the 1981 Israeli air strike against Iraq's nuclear power plant at Osirak. The facility had a nuclear reactor that could have been converted in the future to produce weapons-grade nuclear material, though the International Atomic Energy Agency was monitoring the site to prevent weaponization. But it was no secret that Israel was a prime target for Saddam Hussein's military ambitions. Israel was not willing to wait until Scud missiles that could deliver a nuclear payload were launched. Indeed, it was unwilling to wait until the bombs were even built; it decided to strike when Hussein's nuclear program was still in its development stage. Israel launched its airplanes, crossed into Iraqi territory, and destroyed the reactor in a military action that triggered criticism around the world. Many legal scholars assumed that the attack was illegal because it was not in self-defense. But could one make the claim that Iraq was in the process of attacking Israel because it had promised to destroy Israel and was now building the bombs to make this happen? This argument is provocative, although it is a stretch to pull back our definition of the attack so far that it extends to the time when the bombs are first being manufactured in the munitions factory. This strains credulity. Our definition of an attack must be something broader than troops crossing the border, but something narrower than mere preparations for war such as strategic planning.

So the only way to defend the Osirak strike would be to label it "preemptive" and carve out an extension of the doctrine of legitimate defense that permits preemptive action, even in the face of Article 51's language that an armed attack must occur. Why should we call responses to imminent attack preemptive? As we have noted, the term carries negative associations in the minds of criminal lawyers, and in many international conflicts the popular use of the word echoes its use in domestic criminal law. During the cold war, for example, there was considerable debate about a preemptive strike against the Soviet Union. Both sides stood prepared to launch humanity-threatening nuclear missiles against the other, yet neither side moved to strike first. If one side had feared that the other was preparing for war and decided to save itself by trying to eliminate the missiles of the enemy, it would have considered its action a preemptive strike. It would have thought its use of force was politically necessary, but it could hardly have claimed that its preemptive strikes met the criteria of international law in the United Nations Charter. A preemptive strike during the cold war would be paralleled in the domestic context by Bernard

Goetz's intuition that all the blacks in the subway were dangerous and prepared to attack him. Striking first against the Soviet Union would have been equivalent to Goetz's pulling his Smith & Wesson .38 prior to any threatening action and thereby preempting the use of force. There is little doubt in the criminal law of any country that a preemptive strike based merely on fear is illegal.

Indeed, the term "preemptive strike" *should* carry a connotation of illegality. It falls below the threshold of defensive action against an imminent attack. The U.S. attack on Iraq in March 2003 was clearly preemptive: there was no proof of an imminent threat against the United States or any other country. Even if the White House's evidence of Hussein's weapons arsenal had been accurate (and now we know that it was not), the most the Bush administration could claim was that Hussein was preparing a nuclear force, either to aggress against other countries in the region or to intimidate them to conform to his policies.

It is dismaying, then, that in his first debate against President Bush in the 2004 campaign, Democratic presidential nominee Senator John Kerry said that he subscribed to the doctrine of preemptive war. These are his startling words: "The president always has the right and always has had the right for preemptive strike. That was a great doctrine throughout the cold war."[9]

This is flawed and, frankly, irresponsible rhetoric. Kerry had an opportunity to distance himself from the conception of presidential power that led to the war in Iraq and he failed to do so. His answer to the moderator's inquiry did not mention international law; instead, he introduced the misleading phrase "global test":

> You've got to do [the preemptive use of force] in a way that passes the...global test where your countrymen, your people understand fully why you're doing what you're doing. And you can prove to the world that you did it for legitimate reasons.

This was essentially right, but formulated in a discouragingly confused way, and Kerry then had to defend himself repeatedly against Bush's charge that he wanted to give foreign nations a veto over our use of defensive force. Obviously he did not mean to vest veto power over American defense in the Security Council or any other organization. But he might have avoided this implication altogether by simply citing the Charter, which explicitly recognizes that no nation need surrender its inherent right of self-defense.

However artlessly worded, Kerry said something very important in his use of the phrase "global test." What he meant (and should have said) was that the use of defensive force should be based on public evidence—evidence that the entire world can see. The administration's claim that Hussein's possession of a few aluminum tubes proved his dangerous potential hardly satisfied the global test of persuasive proof. Nor was Secretary of State Colin Powell's presentation to the United Nations more satisfactory. Powell projected a photograph of a trailer and said, essentially, "This is where Saddam is manufacturing chemical weapons."

The best example of public proof was the exemplary presentation made by U.S. Ambassador Adlai Stevenson to the UN General Assembly during the Cuban missile crisis in October 1962. The photographs proved that Castro was installing Soviet-supplied potentially aggressive land-to-land ballistic missiles capable of delivering nuclear bombs to American soil. These photos met the global test. The world community was convinced that Castro was dangerously changing the strategic relationship between the Communist bloc and the United States. Stevenson could confidently dare Soviet Ambassador Andrei Gromyko to provide contrary evidence. He was ready to wait "until Hell freezes over" for the Soviet reply.

As a result of this public demonstration of the danger, President Kennedy's moderate act of self-defense—the blockade of shipping lanes to Cuba—arguably met the criteria of necessary action against an imminent threat.[10] Admittedly, there was no proof that Cuba intended to attack the United States, but it did intend to alter the balance of power between the superpowers. Kennedy's response, coupled with a secret agreement with the Soviets to remove American missiles from Greece and Turkey, contributed to the stabilization of power during the cold war. The Soviets dismantled their missiles in Cuba and the world was made safer for both sides.[11]

Perhaps the terminology of international law should conform more closely to the language of domestic criminal law, which treats the use of force against apparent imminent aggression as a standard case of justifiable self-defense. Also, the word preemptive unnecessarily demotes the status of self-defense against imminent attacks—when, in fact, that is the only kind of self-defense there is. Actual self-defense is based on a response to an imminent threat of harm. Indeed, this scenario is the only case of legitimate and properly defined self-defense.

The danger of our current usage—and this became evident in the 2004 presidential debates—is that we tend to confuse three different kinds of military action: actual self-defense, preemptive war, and preventive war.

In the history of warfare, states have frequently engaged in preventive war when they thought the alliances arrayed against them were becoming too powerful. A classic example is the war of Spanish succession, which broke out in 1700 when it appeared that France might lay claim, by legal means, to the Spanish throne. Any strengthening of the alliance between the major powers of France and Spain threatened both the military and trade interests of countries like England and Holland. When one state threatened to become too powerful too quickly, maintaining the balance of power in Europe meant going to war.

Preventive war is a splendid example of what Clausewitz meant when he described war as "politics by other means."[12] Despite efforts by the United Nations to prohibit the political use of force, the notion of preventive war remains viable—and worthy of discussion—in the post-9/11 period. President Bush's doctrine of preventive war, outlined in a speech at West Point in June 2002, states that the United States "must take the battle to the enemy, disrupt his plans, and confront the worst threats before they emerge."[13] He then added, "In the world we have entered, the only path to safety is the path of action. And this nation will act." This is an extreme application of the principle that the best defense is a good offense. In other words, in the Bush doctrine, the best form of self-defense is preventive aggression.

It would be interesting to apply these categories—actual self-defense, preemptive strikes, and preventive force—to domestic criminal law. We all seem to accept the basic principles of self-defense, and we might also be tempted to use a preemptive strike when we think that someone we fear is preparing an aggressive attack. The first would be certainly legitimate (except in the eyes of some pacifists), and the second might *appear*

TABLE 7.1. Describing the Use of Force

Action	Domestic Law	International Law
1. Response to a shooting in progress	Actual self-defense	Self-defense (armed attack occurring)
2. Using force against someone who is pointing a gun and threatening to shoot	Actual self-defense	Preemptive use of force
3. Using force against someone stockpiling weapons	Preemptive (i.e., illegitimate) use of force	Preventive war (the Bush doctrine)

legitimate to some people, depending on their assumptions about who is the virtuous party and who is the rogue. But it is hard to imagine a legitimate claim of *preventive* individual self-defense. Applied to an interpersonal conflict, the Bush doctrine would generate constant criminal aggression, designed to keep "possible enemies" off guard.

The dissonance between the languages of domestic and of international criminal law is summarized in Table 7.1. It would obviously be in everyone's interest to bring these two spheres of language into harmony.

2. BATTERED WIVES AND BATTERED NATIONS

The requirement of imminence admittedly has its opponents. At one extreme is the United Nations Charter, which, if interpreted literally, would either eliminate the right of self-defense or, as a practical matter, reduce the exercise of that right to the period between the moment enemy missiles are launched and the instant they strike. At the other extreme we find the Model Penal Code, which as early as 1962 sought to water down the required threshold for attacks that could be legitimately resisted. The MPC created a new standard: Force should be permissible whenever the defender reasonably believes that "such force is immediately necessary for the purpose of protecting himself against the use of unlawful force by such other person on the present occasion."[14] Many of the states that reformed their criminal codes in the 1970s and 1980s followed this standard of immediate necessity.[15]

At the time the MPC was drafted, the argument was that certain hypothetical cases could not be properly solved under the standard of imminence. Typically, these were cases in which the threat was slow but inevitable. Suppose a terrorist threatens to implant an undetectable nuclear device that is set to explode in a year. He can be stopped now, but once the device is implanted, it will be too late. Cuba's installation of missiles threatening the United States comes close to this hypothetical situation. In these cases, the attack is not imminent, but the threat is real and ineluctable, and recognizing a right of legitimate defense would seem sensible and appropriate.

The MPC's innovation anticipated a debate that erupted a few decades later about battered women who kill their batterers in their sleep.[16] To make the threat seem serious and inescapable, imagine that before falling asleep the batterer threatened to kill the woman when he woke up. Of course, the woman could run from the house. To counter this objection, the defenders of battered women argued that the female victims suffered from a psychological condition called "learned helplessness."[17] As a result of "battered woman syndrome," they had lost the capacity to take certain precautionary measures. If this syndrome is accepted, then one might well argue that the defendant reasonably believed that killing was "immediately necessary on the present occasion." The MPC standard proved more attractive to battered female defendants than the conventional and traditional standard of imminence. Accordingly, the imminence standard encountered much opposition in the wave of feminist writing that swept the law reviews in the last decades of the twentieth century.

It is no wonder that advocates for battered women were eager to support the MPC standard of immediate necessity. This seemed to accord exactly with the situation faced by many victims of domestic abuse. Although the batterer had inflicted life-threatening physical abuse in the past, the women sometimes killed their abusers while they were sleeping, watching TV, or otherwise engaged in passive recreational activities—a point in time when there was no imminent danger. They chose these moments because they knew that there was more abuse to come and that this particular moment in time represented the only chance to defend themselves from attack. If they waited until their batterer woke up—and the beatings resumed—they would have no reasonable chance of defending themselves. But if they acted now they could save themselves. Their use of defensive force was immediately necessary under the circumstances, at least according to their lawyers.

The immediately necessary standard was also beneficial for another important reason. The earlier portrait of women suffering from battered woman syndrome suggested that they were so abused that they were totally confused and unable to determine when they were really in danger. Consequently, they misread the situation and lived in a constant state of fear. This legal argument emphasized the psychological effects of the syndrome and made the defense sound like an excuse. The women were so psychologically fragile that they should not be held responsible for their actions. But the new standard emphasized that the behavior of the women was rational and based on a calculation that they had one last good chance to protect themselves before their abuser woke up again.[18] This sounds much more like a justification, which accords with the feminist intuition that the actions of the abused woman were justified and lawful, and not the result of psychological infirmity.[19] Tellingly, most victims of the disorder felt justified in their actions and did not run from the scene or attempt to evade capture. They were willing to tell the police what they did—explain their actions, as it were—for the simple reason that they believed that what they had done was justified.

Of course, the success of this defense depends on whether, as a factual matter, the killing of an abusive person in his sleep is really immediately necessary under the circumstances. This depends on the facts of each particular case, the nature and extent of the abuse, and the beliefs of the woman. Regardless of whether these claims are successful, though, the important point is the shift in standard to one of immediate necessity. This innovation of the Model Penal Code represents a fundamental shift in our theory of legitimate defense and could, furthermore, provide a theoretical foundation for a preemptive attack. Just as the battered woman can launch a preemptive attack if it is immediately necessary under the circumstances, can a nation launch a preemptive war if that moment in time represents its last best chance to defend itself? Our view is that if preemptive war is ever justified as an exercise of legitimate defense, it would have to be justified using the Model Penal Code approach. This should be the conceptual framework for our analysis.

Some academics have tried to apply this approach to "battered nations,"[20] although these arguments have not yet filtered down to the level of international lawyers. If they did, the UN might be more sympathetic to the various Israeli actions that reflect the constant urgency of counteracting Palestinian terrorists. In international affairs, the historical trend toward weakening the criterion of imminence derived largely from the fear of nuclear proliferation. No one thought of invading France when de Gaulle acquired his *force de frappe*, but when certain rogue states threatened to join the nuclear club, the situation looked different. When Iraq started to build a nuclear reactor in 1981, allegedly for peaceful purposes, Israel feared that Iraq would use the Osirak reactor to make a bomb and launch an attack against it; Israel sent in its war planes. At the time the United Nations condemned the use of force as unlawful, there was no threat of imminent attack.[21] Israel's official defense was that it was still in a state of war with Iraq, which had never signed an armistice agreement after the Israeli War of Independence—which was fought against Iraqi troops, among others. This might have been true, but there was a de facto armistice, which Israel had unilaterally violated by sending in its planes.

Israel's best argument was that it would be too dangerous to tolerate nuclear weapons in the hands of a sworn enemy. This argument applies the battered woman syndrome defense to international relations, though the accent is on the rogue batterer rather than the battered victim. Israel's allegedly defensive action gave birth to an argument that came home to roost—coincidentally—in the U.S. invasion of Iraq in March 2003, when the Bush administration argued that it had to intervene to prevent Hussein from deploying weapons of mass destruction. Though we now know that this suspicion was false, we must assume for purposes of analysis that the administration had reasonable grounds for its fears.[22] Does not the possibility of WMD pose a serious challenge to the requirement of imminence as the threshold of a legitimate defensive response?

One might argue that, under the MPC standard, the United States thought it was immediately necessary to invade Iraq, but one would be hard put to apply the proviso "on the present occasion" to the prolonged ten-year period in which the United Nations imposed sanctions on Iraq and conducted inspections to control its supply of weapons. Nothing in particular had happened to support the decision to invade on March 20, 2003—or indeed, in 2003 rather than 2004.

In Israel's case, timing was obviously more relevant. Israel knew Hussein was building a reactor. Once the reactor was in place, the Israelis would not know whether plans were under way to exploit the resulting enriched uranium to build a bomb. They might have plausibly argued that, though the threat was not imminent, the use of force was, in the language of the Model Penal Code, "immediately necessary on the present occasion."

The problem with this argument is that it leads us to base our determination of necessity on efficiency. Actions appear to be necessary if they are based on a reasonable judgment of the costs incurred, relative to the benefit of undertaking the action and counteracting the threat. In cases of individual self-defense, the calculations are simpler. It is reasonable to protect your life when faced with a potentially deadly attack, even if the cost is that the attacker will die, because the value of your life is high. No one doubts this. If you apply the cost/benefit approach to international disputes, however, the calculations become somewhat more complicated. The United States would have to claim that the danger posed by a rogue state acquiring WMD is so great that violence is justified to counter the risk. The number of deaths now would be smaller than the future deaths caused by the rogue state. (Indeed, the U.S. argument would have evaporated if it were forced to admit that invading Iraq caused more civilian deaths than would have been the case if Hussein had retained his weapons and used them against his enemies.) Unfortunately, this creates a standard that anyone can apply in any case in a self-interested manner. A coalition of Islamic states fearful of American intervention might justify bombing the United States based on the same perception of risks and benefits.

The facts of life in the nuclear age have forced us to concede that the standard of imminence is contestable, but it is hard to know what should replace it.[23] Cost/benefit analysis supports intervention when the perceived danger is great enough. In the view of the Bush administration's policies, there was little doubt that the danger justified the invasion of a rogue state. However, if we believe that in the future the international community should avoid the sort of mistakes we have witnessed in Iraq, perhaps we should reconsider whether the imminence standard should be defended against those who would ease the threshold of legitimate defense.

The Israeli attack against the Osirak reactor and the U.S. invasion of Iraq highlight two major problems that have received virtually no attention in the literature of international law: whether it is possible to maintain the rule of law without adhering to the principle of reciprocity, and whether the legitimate use of force must meet a "global test," in the sense that it must be based on publicly observable facts. Each of these elementary requirements of the rule of law requires comment.

First, on the issue of reciprocity. To put it colloquially, what's good for the goose must be good for the gander. There is no way that Israel can claim a right to strike the Osirak reactor without recognizing the reciprocal right of Iraq to attack Israel's reactors in Dimona. If one is a threat to peace, then so is the other. The same is true of the U.S. threat to invade Iraq, which became apparent—and indeed, imminent—as soon as Congress passed its joint resolution in October 2002 authorizing the president to use military force against Hussein's regime. As many observers have noted since March 2003, the rogue states of the world—most notably Iran and North Korea—now have greater incentives than ever to acquire and deploy nuclear weapons as the only way to deter aggressive action by the United States.

It is tempting to slip into the pattern of distinguishing between the supposedly peace-loving democracies and rogue states, and this temptation accounts for our failure to think more critically about the issue of reciprocity in self-defense.[24] Kant thought that we could distinguish ourselves from "unjust enemies."[25] We take for granted that the United States, the United Kingdom, and Israel can rightly protect themselves by military means because they are peace-loving democracies and that rogue states must not acquire nuclear reactors or nuclear weapons because they are presumed to have aggressive intentions.

It might not be obvious to many policymakers that classifying states in this way runs afoul of a basic value of the rule of law. The idiom of reciprocity does not come easily to the lips of international lawyers, yet equivalent concepts play a decisive role in domestic reflections about the rule of law. Indeed, it is hard to find serious jurisprudential reflection on the latter topic without emphasis on general, universally applied rules.[26] If such principles govern the rule of law in domestic law, they should also apply in the international arena as an aspiration of the law of nations. The rules of any domestic or international legal system must treat everyone equally and be applied generally. There is no way to build in a distinction between "good guys" and "bad guys," and making these distinctions in international law recalls the scheme advanced by Carl Schmitt to distinguish between friends and enemies.[27] Though many states now unfortunately have one criminal law for citizens (friends) and another for recidivists and terrorists (enemies), you cannot replicate this practice at the level of international law.[28] States may have the power to enact repressive internal laws and thereby institutionalize discriminatory practices designed for enemies, but the international community must live or die on the basis of principles that are applicable to *all* states—precisely as the UN Charter prescribes. International law is based on the premise that states are free and equal persons under the law.[29] The community of democratic states does not have the power to designate some nations insiders and others outsiders.

Of course, it might be possible to water down the standard of imminence and to apply the resulting weaker standard across the board, but this poses additional problems. Suppose that Israel were justified in attacking the Osirak reactor; would Iraq also be justified in defending its territory against attack? One would think so. Israel would certainly defend Dimona against incoming Iraqi warplanes. Thus both sides might lay claim to their inherent right of self-defense. If applied at the domestic level, this easy recognition of defensive rights would strike criminal lawyers as a contradiction: it cannot be the case that an abused woman is justified in attacking her sleeping batterer, but if he wakes up under a falling knife, he cannot justifiably act to avoid the assault. This creates exactly the kind of incoherence that bothered Kant. If we universalized this maxim of preemptive intervention at the international level, the result would be a world of total violence, devoid of stability and full of aggression. It is precisely for this reason that we cannot logically, upon pain of contradiction, advocate for a doctrine of preemptive war and simultaneously demand that our enemies refrain from it.

If we assume that in any particular conflict both sides are of equal moral standing and only one side may claim self-defense against the other, then we need a rigorous standard for recognizing legitimate defense—more rigorous than the MPC criterion of immediate necessity or the Bush administration's implicit appeal to cost/benefit analysis. It must be a standard that applies equally to both sides. Each side must be classified as lawful or unlawful, defender or aggressor, on the basis of what it does. The appeal of imminence is precisely that it provides a nearly foolproof standard for distinguishing between the aggressor and the defender. The defender can justifiably use force, whereas the aggressor, if he kills, is guilty of homicide, or at least unlawful aggression. This is the appeal of requiring the defending party to wait until the point of imminent attack. Waiting ensures a stronger case that only the defender is acting in legitimate defense.

The standard of imminence works properly when the threatened aggression is manifested in publicly observable facts. When the facts are laid bare—when the threatened attack is manifested in troop movements, missile deployments, and the like—the world will know which side to support in the struggle. Outside powers are permitted only to support the defender against aggression, and therefore there must be some public test of aggression. When there is persistent disagreement about the evidence, as there was in the case of Iraq, the invading powers might believe their action is justified, but a large number of states, particularly those who identify with the victim state, will treat the invaders as aggressors and criminals.

The principle of publicity is critical in a self-administering system of law. There is no court to determine the facts underlying international legal conflicts; there is no authority but the eyes of the world to assess whether the United States had sufficient evidence to warrant its claim of dangerous and deployable WMD in Iraq.

These two fundamental principles—reciprocity and publicity—have gotten lost in the debate over whether Saddam Hussein really was the worst villain in the world. Regime change is not a justification for aggression. There might be some who think that the aim of spreading freedom and democracy justifies deposing foreign dictators and staging free elections.[30] But it is hard to avoid the conclusion that these righteous goals are but the modern secular analogue of the just wars of the Christian tradition. Just causes and a righteous purpose hardly suffice to establish a fair, reciprocal rule that all states can live by. However, if the relevant criterion is not just but *lawful* war, a good case remains for retaining the criterion of imminent attack. This conventional standard of domestic law best realizes the principles of both reciprocity and publicity. Reciprocity ensures that both sides are treated equally and fairly under the law, and publicity guarantees a standard that can be fairly administered by the states themselves, without the urgent intervention of a court to determine who is right and who is wrong.

Reciprocity is so misunderstood in our current environment that it leads to a totally asymmetrical view of warfare. For example, the Bush administration maintains the right to torture suspected terrorists, either directly at CIA detention facilities overseas or indirectly by extraordinary renditions to friendly states that practice torture.[31] However, the key to the Bush doctrine is that torture is justified against *terrorists* and that our right to torture suspected terrorists for information does not entail a reciprocal right for them to torture us. Furthermore, the Bush doctrine could even appeal to legitimate defense as a justification for torture: to defend others from terrorist attacks, government officials must occasionally use physical violence against detainees.[32] But all of these arguments for torture, regardless of how they are framed, suffer from the same defect. We deny terrorists the right to be free from torture, but we are not willing to forgo this right for ourselves; if American soldiers were captured abroad and tortured, we would surely complain. This lack of reciprocity flies in the face of a coherent account of preemption. If we can use torture to prevent armed attacks, then our enemies can use torture against us as well.

It should be noted that terrorists, too, as a group, are guilty of ignoring reciprocity. Terrorists seek political objectives by killing innocent civilians. However, terrorists do not concede that it would be right for us to kill their families or the innocent civilians on whose behalf the terrorists are fighting. Indeed, they assume that our status as imperialists or infidels or some other moral failing justifies a lower level of protection in warfare. Because of who or what we are, we do not deserve to live. But this conclusion is asymmetrical. Similar acts of desperation on our part would draw their criticism—indeed, it would be taken as more evidence of our moral failings—just as the United States would criticize any foreign group that dared use torture against our troops. Although none of this entails that the torturer and the terrorist are the same (or are equally guilty), it does suggest that both depart from the normal conduct of warfare through the same conceptual deficiency: lack of reciprocity.

3. MISTAKEN BUT REASONABLE BELIEFS: WEAPONS OF MASS DESTRUCTION IN IRAQ

The invasion of Iraq also highlights the problem of mistaken beliefs, which is an issue distinct from imminence. Colin Powell gave a long presentation to the United Nations explaining the evidence that Hussein was still actively pursuing weapons of mass destruction. Powell showed evidence of chemical weapons manufacturing and suggested that Hussein was actively pursuing a nuclear device through the black market. As it turns out, of course, this presentation was completely false. Hussein had no facilities for manufacturing or researching chemical weapons and had apparently abandoned his effort to produce major amounts of them. We do not know why. Perhaps the pressure from the UN weapons inspections made the program unworkable, or maybe it was too expensive, or maybe he just preferred to let the international community believe that he had the weapons, when in reality he did not. In any event, the evidence presented by Powell was false. After the invasion, a special military unit was specifically tasked with scouring the Iraqi countryside looking for WMD and they never found them. After an embarrassingly long time, they just gave up the search altogether.

The first issue is whether Powell and Bush were sincere in their beliefs. Many have asserted that the evidence was skewed by the intelligence community after intense political pressure from White House officials, who were accused of interfering with the supposedly objective process of intelligence gathering and risk assessment. If this is indeed the case, then the invasion was clearly unjustified because then Bush's whole argument for preventive war collapses. How can the international community determine if a mistaken belief was sincere or was used as a pretext to justify military invasion? The Bush administration could say that Hussein had WMD, not really believe it, launch the attack, and then issue a mea culpa afterward for their "reasonable" mistake. This would turn the doctrine of preventive war into a farce. The only solution is to require that the evidence be, in some manner, publicly available to the international community (say, through the General Assembly or the Security Council) and not relegated to the president's hip pocket.

But let us assume, for the sake of argument, that the beliefs were sincere and analyze whether the justification still holds if the sincere belief turns out to be mistaken. If the presence of WMD was the justification for this preventive war, what happens now that we know that we were wrong about the threat? There is always a problem of mistaken beliefs in a theory of legitimate defense, and the issue has been long debated in the domestic theory of criminal law. The whole issue in the *Goetz* case was how we should evaluate Goetz's mistaken belief that he was about to be attacked on the subway, and whether or not his belief was reasonable and under what standard. But the problem is even more acute in the case of preventive war, because preparations for war are often conducted in secret, away from the international public, unlike an armed attack that is available for all to see. There were witnesses to the Goetz shooting, so we know exactly what was said and how it went down. But Hussein's weapons programs were conducted in secret and, at times, hidden even from the eyes of UN weapons inspectors. Preparations such as these are often discovered by the intelligence community, but intelligence can be misinterpreted, faulty, or manipulated.

So the first question is whether a mistaken but reasonable belief bars the right to go to preventive war. The Bush administration mistakenly believed that Saddam Hussein had WMD capabilities. In criminal law, as long as the mistaken belief was objectively reasonable at the time, then the doctrine of legitimate defense would apply, and at first glance there is no reason to think there should be any difference in international law. The real question is whether we should view this defense as a justification or as an excuse. One might be confused as to why this distinction is important, especially in international law. Why does it matter whether we view the Bush administration's mistaken views about Hussein's WMD as a justification or an excuse? The answer, of course, lies in Iraq's right of resistance to the invasion. If a preventive invasion is justified, then the conduct is lawful and not a case of aggression. Iraq therefore has no right to resist, and its use of defensive force is illegal. This follows the model of domestic self-defense.[33] If you are justified in using legitimate defense against your attacker, your attacker cannot legally use force to fend off your counterattack. He is the original aggressor, not you. If, on the other hand, the preventive invasion is based on a mistake, and the target country was not harboring weapons of mass destruction or preparing for aggressive war, then it is clear that the target country has the right to fend off the invasion, consistent with its own rights of legitimate defense. Why should it be forced to give up its right to self-defense just because the invaders had a mistaken belief? Again, this way of analyzing the situation follows the model of domestic self-defense. If a neighbor mistakenly thinks you have a gun and must be stopped, you have every right to defend yourself against him, even though he might have an excuse and not go to jail because of his reasonable mistake.

International lawyers are less than sensitive to this distinction, although a few scholars with a background in criminal law have mentioned the issue.[34] But no one has fully recognized the degree to which excuses may be important in international law. The international community is uncertain about how to characterize the U.S. invasion of Iraq in terms familiar with international law. The answer they are looking for is to be found in the language of criminal law. The U.S. invasion was not legally justified, and the Iraqis were justified in resisting the attack, although the United States may have an excuse based on its mistaken beliefs about chemical and biological weapons.

What does it mean for the United States to have an excuse in this situation? We discussed this problem in chapter 5. It seems strange to talk about excusing a nation, and this term is completely foreign to the modern language of international law. It seems odd to attribute to a nation the kind of mental states that go along with an excuse, but this problem is a general one having to do with our treatment of states as persons capable of being held responsible by the international community. If we can hold states responsible for their actions, there seems to be no reason why we cannot attribute both excuses and justifications to their actions. As noted in chapter 3, international law in previous centuries was once more comfortable with the language of excuses, and resurrecting this language presents an intriguing possibility. The current situation in Iraq demands clarification. Presenting this excuse could be relevant at a case brought against an aggressor at the ICJ (although the United States of course refuses to participate in ICJ proceedings). But the theoretical issue would be important in other international conflicts where the international community decides that punitive action must be taken against another country.

The final issue is whether an *unreasonable* mistake about weapons of mass destruction could be offered as a defense—either an excuse *or* a justification—for a preventive war. Some might argue that the United States was negligent in its collection and interpretation of intelligence in the run-up to the war. Administration officials allegedly analyzed the information with a particular conclusion in mind and refused to look for information that might disprove their hypothesis. This kind of selective intelligence gathering, if true, might be regarded as a form of negligence. If the United States was indeed negligent in gathering its information, one might well want to conclude that its mistaken beliefs about WMD were sincere but *unreasonable*.

As a policy matter, unreasonable beliefs are disfavored in criminal law. If Goetz's mistake was unreasonable, the New York State Court of Appeals would not excuse or justify his use of force. Of course, the issue of unreasonableness hinges on whether we use a subjective or objective standard of reasonableness. The court of appeals answered in *Goetz* that the appropriate standard was objective and based on community standards.

The same should hold true in international law, for the simple reason that states should be excused from their mistakes only if those mistakes were reasonable, not just in their eyes, but also by reference to an objective standard. The question is not just whether the United States believed that Hussein had weapons of mass destruction; the issue is whether it was objectively reasonable for any country to believe, with the information available, that he posed a danger with these weapons.

This underscores the importance of the publicity constraint discussed earlier. Evidence for a preventive war must be publicly available to the world community before the invasion occurs. We have to be careful about former Secretary of Defense Donald Rumsfeld's argument that "the absence of evidence is not evidence of absence."[35] If the evidence for an invasion can be gathered only after the invasion itself, the publicity constraint is not met. But if the evidence is available and the international community can see that it is objectively reasonable, then the exercise of legitimate defense in the form of a preventive war can proceed. Again, the Cuban missile crisis is the paradigmatic example: we had the satellite images and published them for the whole world to see. The publicity constraint will therefore limit the number of mistakes that might happen and limit the need to even resort to mistaken beliefs as an excuse.

4.   DISTURBING THE BALANCE OF POWER

A final issue remains. The other rationale for launching a preventive war is to redress a change in the balance of power in a region. Is a mere shift in the balance of power sufficient to justify an attack under our standards of imminence, reciprocity, and publicity? Kant felt it was sufficient on the theory that an alteration of the balance of power between states was a significant form of aggression. A state need not watch as its regional influence is diminished to the point of being so weakened that it can no longer defend itself. Two case studies will be considered.

During the Cuban missile crisis, Kennedy initiated a blockade against Cuba because Khrushchev's missiles in Cuba altered the balance of power in the Americas. For the first time there was Soviet influence within striking distance of North America and directly off the coast of Florida. The nuclear missiles were capable of striking the eastern seaboard. This frightened the American public, although certainly the U.S. government encouraged these fears to justify the blockade and gain both domestic and international support for it. At the time, many international legal scholars did not consider the shift in the balance of power in the region to be sufficient to justify military action, and the blockade was criticized. However, it would seem likely that the shift in power was so sufficient—and threatening—that Kant's argument holds some sway. The passage of time has been kind to Kennedy's decision, and the move is credited with preventing a nuclear war. The Soviet move was clearly provocative and increased Soviet strategic interests in the region at the expense of American security.

But contrast the blockade with another, more infamous preemptive attack: Pearl Harbor. The argument for preventive war to redress a shift in the balance of power looks suspicious when we consider it from the point of view of our enemies. The Japanese attacked Pearl Harbor because a buildup of U.S. naval forces in the Pacific was changing the balance of power in the region. The United States was attempting to use its naval forces to project its influence in the Pacific to counter Japanese expansionism and aggression, in effect challenging Japan's status as the preeminent imperial power in the Pacific region. Of course, the U.S. ships were not engaging in military assaults, but they were undeniably in the region. From the point of view of the United States, its naval capabilities in the Pacific did not seem threatening, provocative, or aggressive. But to the Japanese, the increase in naval capabilities put the U.S. military within striking distance of Japan and challenged its status as the region's superpower that it had acquired after incursions into Manchuria in 1931 and a larger invasion of China in 1937. In response to what the Japanese viewed as an inevitable naval confrontation, Japan struck first. The Pearl Harbor attack—on a military installation, no less—was regarded internationally as illegal and an unprovoked act of war. In the United States it was a day that would "live in infamy."

But if the balance of power argument is accepted, do we need to reevaluate the legal status of Pearl Harbor? This provokes an uncomfortable tension in our intuition. If we accept the argument that a change in the balance of power may justify a preventive war, then we would be forced to admit that this argument is equally valid for our enemies, not just our friends. This would be especially problematic for a country such as the United States that is continually altering the balance of power in its favor by further projecting its military influence around the globe. If our enemies have the right to launch preventive wars against us each time we change the balance of power in our favor, the United States will be on the receiving end of a constant barrage of attacks. Because the United States is always changing the balance of power in its favor in all regions of the globe, dozens of our enemies have the right to attack us to return the balance of power to their favor. This is a recipe for world war, and Kant's theory no longer looks tenable.

The problem with balance of power arguments is that they violate the reciprocal criterion that we discussed earlier. The Bush administration likes to argue that a shift in the balance of power justifies our use of preemption, but we do not extend the same privilege to rogue states. Rogue states are on the *receiving* end of this argument, just as they are on the receiving end of our bombs. But once again this opens up a nonreciprocal treatment of states within the international system, which is intolerable. Just as your use of legitimate defense on the street does not depend on your station in life, so, too, a nation's use of legitimate defense in the international sphere should not depend on its status within the international community. All that matters, in both cases, is how the two opponents stand in relation to each other. The doctrine of legitimate defense must be the same for all.

8

THE COLLECTIVE DIMENSION OF WAR

By now it has become clear that self-defense comes into play at three distinct levels of legal argument. The standard point of reference is self-defense in domestic legal systems, against domestic crimes, and then there is the newly defined standard of self-defense against war crimes (and possibly other offenses) under the Rome Statute. These two dimensions of self-defense have roughly similar contours, as they are both invoked by individuals to justify their apparent crimes of violence against other individuals or groups. In between these two forms of individual defense lies the commonplace but underexplored phenomenon of collective self-defense by states against other states (and nations against other nations). This is the inherent right of legitimate defense recognized in Article 51 of the United Nations Charter and discussed in previous chapters.

The term collective self-defense can be confusing because Article 51 uses the same term to refer to one state's going to the aid of another. For the drafters of Article 51, individual self-defense means that the state attacked defends itself. Collective self-defense means that two or more states together fend off an attack. The better term for this situation might be "joint self-defense," to indicate that two states, treated as individual actors, are joining forces to thwart a common enemy. An analogous sense of collective defense comes to the surface most clearly when states are obligated by mutual defense treaties such as NATO or the now-defunct

SEATO to come to each other's aid in the event of an attack. For this chapter, however, we should put aside these meanings of collective in the sense in which it is used both in the UN Charter and in regional security pacts.

For our purposes, the relevant sense of collective lies in the idea of group action, both for purposes of going to war and for purposes of self-defense against aggression. When one nation attacks another, its aggression is collective; when the attacked nation defends itself, it also acts collectively. When the collective acts, the whole is greater than the sum of its parts. The nation and its army enjoy an existence of their own. The implications of collective action in the law of war are systematic and deep, as has been clear since the beginning of this book. If Adolph, acting in a military unit of the state Southland, attacks Belsky and other members of the same military unit representing Northland, this is an act of aggression by Southland against Northland. A hundred miles away, Northlander Chopin's commander hears of the attack and orders his unit, including Chopin, to attack Southlander Detlef and his unit. Chopin kills Detlef. The way the aggression of Adolph (A) against Belsky (B) justifies the response by Chopin (C) against Detlef (D) is illustrated in Table 8.1.

The only way to comprehend the justifiability of C's action is to embed it in a set of collective responses by Southland and Northland. B is attacked, and therefore C is entitled to respond. This makes no sense without a sense of collective identity that links the injury of one to the other's right of response. The same is true on the other side. D must suffer only because as a soldier his fate is bound to the consequences of A's aggression against B.

Benedict Anderson famously used the label "imagined communities" to describe the links between A and D, parallel to those between B and C.[1] The two may never have met; they may not even know of each other's existence. But, in the typical case, they speak the same language, read the same newspapers, pray with the same translation of the Bible,[2] and revere the same set of national symbols. Their communities and their

TABLE 8.1.

Army of Southland	Army of Northland
A →	→ B
D ←	← C

identification with each other are not made up or fictitious, but a product of imagination in the positive sense.

The important implication of collective thinking in the law of war is that neither A nor D is guilty of a crime. There would be no defense to their use of force under domestic legal systems, but here their injuring or even killing others is absorbed into the actions of the collective by the law of war. If anyone is guilty, the collective is guilty. A's collective, the nation or state of Southland, might be guilty of a conspiracy to commit aggression. This was the crime charged against the German general staff, invented and punished for the first time at Nuremberg. D's collective could be guilty of the same crime if the collective action of Northland is not classified as self-defense.

There are many puzzles in these claims. How is it possible that a soldier's killing can be absorbed into the action of the collective and no longer count as a punishable homicide? In the conventional language of criminal theory, is A's action either justified or excused? Or is it possible that an enemy soldier is not a legally protected interest—something like an outlaw or an animal, the killing of which requires neither justification nor excuse? The latter is probably the traditional law of war. The soldier is so completely identified with the collective that only the collective commits and suffers aggression.[3] Francis Lieber formulated this view in Article 20 of his classic code:

> Public war is a state of armed hostility between sovereign nations
> or governments. It is a law and requisite of civilized existence
> that men live in political, continuous societies, forming organized
> units, called states or nations, whose constituents bear, enjoy, and
> suffer, advance and retrograde together, in peace and in war.[4]

But since the invention of war crimes—at least as early as the 1899 Hague Conventions—it has no longer been tenable to think of enemy soldiers as part of the collective mass, capable of being neither offender nor victim. The Hague Conventions strictly prohibit killing treacherously or using poison.[5] Admittedly, in 1907 only states were responsible for these actions, but these prohibitions against certain forms of warfare are carried forward, word for word, in the Rome Statute, which imposes individual criminal liability. The victim of a treacherous or poisonous killing was presumably always the individual soldier whose life was taken by these means.

When a soldier kills by these strictly forbidden means, he or she emerges from the mass and becomes an individual again. In ordinary fighting, then, we might think of the soldier as part of the collective; in the case of criminal aberrations from the ordinary, however, we must treat the soldier as an individual, accountable for his or her own crime. The problem of drawing the line between the collective and the individual is illustrated by the following two hypothetical takes on war and self-defense.

1. THE CASE OF THE POLISH FARMER

Imagine German troops marching toward Warsaw after the Polish government has surrendered and ordered the army to lay down its arms. Suppose a farmer, standing by himself in the fields, sees the troops marching down the road in formation. He rushes to his barn, takes his rifle up to a window in the attic, and shoots at them as they pass. He kills three soldiers.[6] Does the legitimating effect of warfare encompass these killings, which would otherwise be murder under both Polish law and German martial law?

One naturally has some sympathy for the Polish farmer defending his homeland. At first glance it would seem that he has the right to defend his country in any way he can against the unlawful aggression of the Nazis. But these sympathies find little warrant in international law.[7] The reason lies in the principle of reciprocity, which governs the entire law of armed conflict. Although this concept was introduced in previous chapters, we have not yet fully charted its implications.

Note first that the Polish farmer has no claim of individual self-defense. The German soldiers are not attacking or threatening to attack him personally. The defenders of the Warsaw Ghetto were in a different situation, for they were subject to the constant threat of death and deportation. Also, concentration camp prisoners faced a constant threat of *imminent* death at any moment, and an attack against an enemy guard could be justified on the grounds of individual self-defense. But there is a difficult problem in this context of imminence, a concept discussed at length in previous chapters. Is every sleeping German soldier fair game, even for Jews fearful of genocidal liquidation? Would the situation be different for the Polish farmer if he were Jewish? One might, for example, argue that any Jew in Poland in close proximity to Nazi troops was in danger of being arrested, deported, and gassed at Auschwitz. Furthermore, once a Jew is arrested by Nazi troops, his possibility for escaping, or defending himself with force, was minimal. The only chance for survival would be to avoid arrest in the first place. Under this theory, the shooting might be "immediately necessary," in the words of the Model Penal Code, if not necessarily justified by virtue of imminence. Does this mean that every

Jew could claim individual self-defense as a justification for shooting a passing German soldier? This is a coherent and provocative argument, although it depends entirely on the facts on the ground at that time.

Furthermore, we rejected the standard of immediate necessity in chapter 7 during our discussion of preemptive war and argued in favor of retaining a renewed standard of imminence, based on the principle of reciprocity; that is, so-called rogue nations should be subject to the exact same standards as so-called lawful nations. In the case of the Polish farmer, we have the individual version of this collective dilemma. If we allow the Polish farmer to kill the German soldier, then reciprocity demands that we must allow the German civilian to kill and ambush an American GI in Germany. Should it matter that in one case the soldier is participating in an illegal war, and in the other the solider is fighting a lawful one? To answer this question, we must analyze the idea of reciprocity as it applies to the laws of warfare.

Let us assume for the moment that the Polish farmer is not Jewish. The German soldiers marching down the road present no risk to the farmer's personal security. If the farmer is entitled to defend his country, it is only by virtue of his membership in the Polish collective exercising resistance against the German invasion—a resistance that lingered even after the official surrender. But the right to fight as part of a collective entails an exposure to risk. Suppose the German soldiers, in a military action, killed the farmer as they passed by. They would clearly be guilty of the war crime of intentionally killing civilians, a breach of the Hague Conventions. If soldiers today engaged in similar killings as part of "a plan or a policy," they would be individually guilty under the Rome Statute.[8]

The protection affirmed by the Geneva Conventions for civilians entails reciprocal duties on their part: they are not entitled to think of themselves as free agents acting on behalf of their country's army. This would be true whether or not the government had already surrendered. Because the farmer is acting alone, independently of the army, the case falls outside the realm of collective activity that defines the law of war and reverts to a case that should be tried under domestic law.

This case is so difficult because we assume that the German invasion of Poland is illegitimate. Our sympathies are on the side of the Polish farmer, yet these sympathies have no bearing on the legal analysis of the case. The architectonic assumption of international law is that the right to go to war (jus ad bellum) has no bearing on the law applied in the course of war (jus in bello). The correct result under international law—as hard as it might be to swallow—is that the law of collective self-defense does not apply. The farmer is guilty of murder under domestic law.

The critical point is that the farmer, acting alone—or even a group of partisans acting alone, however appealing their cause—cannot claim the rights of warfare, including the right of collective self-defense. This is the critical line between terrorism and warfare. Terrorists like Timothy McVeigh cannot claim the rights of war for the simple reason that they are not engaged in an armed conflict between organized military forces. The crucial point in this argument is demarcating the boundary between collective actions covered by the law of war and individual crimes punishable under domestic law. The Polish farmer falls on the side of individual action governed by domestic law.

This analysis has a direct bearing on the claims of Palestinian terrorists of a right to engage in treacherous and deceptive attacks on Israeli civilians, settlers on the West Bank, and Israeli soldiers both in Israel proper and in the occupied territories. If Palestinians could articulate a reciprocal exposure to risk, they might have a claim, but so far no defender of the Palestinian cause has been willing to state who is properly and legitimately at risk as a result of resistance against the settlers or the Israeli army. No one is willing to say that the entire civilian population that applauds and supports suicide bombings is subject to reciprocal military action. Nor is any defender of the Palestinian cause willing to allow Israeli intelligence to identify the potential offenders against Israeli security—offenders who would be treated like combatants, subject to being killed in military action.

The question of targeted assassinations is another instantiation of the same problem. A recent Israeli Supreme Court decision addressed the issue of assassinations of suspected terrorists by Israeli Defense Forces in the West Bank and Gaza Strip.[9] If we label the terrorists as criminals, they are subject to arrest but are not legitimate targets of assassination. But if we think of the terrorists as combatants, they are subject to assassination under the laws of war just like any enemy soldier. Justice Barak noted that terrorists are legitimate targets when they bear arms, but he concluded that they cease to be legitimate targets once they put down their weapons. In other words, a terrorist is a combatant only when he is acting like one; otherwise, he retains the protected status of the civilian.[10] This creates what amounts to a 9-to-5 rule. During the day, when they work as terrorists, they are subject to military attack as soldiers, but at night, when they go home for dinner, they must be arrested as criminals. The next question is whether Barak and the Israeli Supreme Court would be willing to accept the full consequences of their view. If the terrorists are combatants, then launching strikes against them should open the Israeli Defense Forces to a reciprocal risk; they, too, become legitimate targets. It is doubtful, however, that Barak or his colleagues on the Israeli high court would be willing to recognize a Palestinian right of self-defense on behalf of targeted Palestinian terrorists. This is proof of the breakdown of reciprocity.

Reciprocity is a demanding standard. Those who resist occupation do not accept it, but in the Palestinian case the Israelis themselves also resist the principle of reciprocity. They engage in targeted assassinations of Palestinian terrorist leaders, even though the targets might be praying in their home or sleeping in their bed.[11] They claim that these terrorist leaders are combatants and therefore subject to the law of war, but they are not willing to accord them the rights of POWs. The law of war is rife with contradictions. Fewer and fewer of those engaged in quasi-warfare today are willing to be bound, consistently, by the basic principles of honorable and reciprocal warfare.

The magic formula for those who want to have their cake and eat it, too—to attack civilians as though they were combatants but not to pay the price under the law of war—is the nefarious idea of the unlawful combatant. As mentioned in the introduction, the U.S. Supreme Court invented this term in 1942 to explain why a military tribunal could prosecute eight German saboteurs instead of treating them as prisoners of war.[12] The concept lay quiescent until 9/11. Since then the Bush administration has pointed to the concept to explain its posture of nonreciprocal warfare. When combatants are unlawful, the argument goes, they are subject to the burdens of combatancy (they can be killed), but they have no reciprocal rights. This way of thinking—deeply embedded in the current mentality of the war on terrorism—exploits the conceptual gap between war and crime. The phrase unlawful combatant as used today combines the aspect of unlawful from the law of crime and the concept of combatant from the law of war. For those thus labeled, it is the worst of all possible worlds.

For those willing to reject the principle of reciprocity to support causes like those of the hypothetical Polish farmer or the real-life suicide bombers in Iraq and elsewhere, some reflection might be in order. States can play the same game of rejecting reciprocity in fighting their war on terror. They can target civilians by calling them unlawful combatants and ignore the rights that all combatants enjoy under the Geneva Conventions.[13]

2. THE CASE OF THE AMERICAN OFFICER IN IRAQ

Our general problem is staking out the boundary between collective and individual action. In the example of the Polish farmer, the appeal to collective action offered a way of defending acts that would otherwise be criminal. In some situations, the frontier between individual and collective action determines not whether the suspect is guilty or innocent, but the legal regime under which the defendant is guilty and what the punishment should be. A good example is a hypothetical case of an American officer in Iraq who falls in love with an Iraqi woman. When he is off duty and wearing civilian clothes, he enters a restaurant frequented by local people. There he encounters the brother of the woman he loves. They exchange words, the brother insults him, and a fight breaks out. The fight escalates, and the American pulls his service revolver and kills the Iraqi. Under the general principle of territorial jurisdiction, Iraqi law should apply. The problem is whether the killing is also a war crime and subject to prosecution at the International Criminal Court.

To focus on the interesting substantive question, let us concentrate for a moment on the jurisdictional issue. As of this writing, neither Iraq nor the United States has ratified the Rome Statute, but that would not prevent prosecution in The Hague if the Iraqi government felt that, due to American pressure, it was "unable" to carry out an investigation and prosecution.[14] By filing a declaration in The Hague, the Iraqi government, as the government of the territory where the crime was allegedly committed, could accept jurisdiction of the court "with respect to the crime in question."[15] Thus, contrary to popular opinion and propaganda disseminated by opponents of the court, the failure of the United States to ratify the Rome Statute does not preclude the prosecution of American soldiers for war crimes committed abroad.

With the jurisdictional issue laid to rest, we can turn to the substantive issue of whether the American officer's killing the Iraqi citizen under these circumstances would constitute the war crime of "wilfully killing" a civilian protected by the Geneva Conventions.[16] The international law of willful killing is stunningly undeveloped. The term willful is not even translatable into the other official languages of the Rome Statute (it is typically rendered as "intentional").[17] Even under the English version we cannot be sure whether a killing provoked by insults or resulting from a love affair would be the kind of heinous homicide signaled by the word willful. Let us suppose the killing was willful. That still does not settle the issue of substantive liability.

The Rome Statute imposes several filters against the prosecution of routine crimes committed by individuals. The homicide must be part of a "plan or policy" or constitute "part of a large-scale commission of such crimes."[18] It must be one of the "most serious crimes of concern to the international community as a whole" and be of "sufficient gravity to justify further action by the Court."[19] The facts that bear on this assessment are the status of the defendant as a military officer and the fact that he used his military weapon. It is also relevant, however, that the crime occurred when he was off duty, wearing civilian clothes, and eating in a local restaurant. The basic question is whether the case is so far removed from the actions of the military that the killing should be qualified as a purely private transgression, not subject to prosecution as a war crime.

The closest analogue to this problem in U.S. law is determining whether an action by a state employee, after hours or during his vacation period, is attributable to the state government. In one case, a teacher engaged in alleged sexual harassment while he was working at an overnight camp during the summer break. The court decided that this was not a violation that the state could be responsible for. The offensive action did not occur "under color of state law."[20] Similarly, if an off-duty police officer commits a rape, this action is not usually the action of the state. The Due Process Clause of the Fourteenth Amendment does not apply to this and other purely personal crimes. The useful idea of acting under color of law captures the distinction between the private and the public, between the isolated acts of individuals and the actions of the collective called the state. In the case of the American officer in Iraq, one might say the issue is whether he is acting "under color of war." That is the proper way to approach the boundary between violations of domestic law and crimes against the law of war.

Admittedly, some thoughtful advocates of human rights argue that every crime committed by an occupying soldier against a civilian occurs under the color of war and constitutes a war crime.[21] This would include every homicide, every theft, every rape, and, in the language of the statute, every "outrage upon personal dignity,"[22] however unrelated these acts might be to military operations and the war effort. This view is hard to accept. It seems far more consistent with the spirit of the Geneva Conventions and the Rome Statute to require that these acts be carried out in the name of the military—in the name, as it were, of the occupying army.[23] Though the point is controversial, the better view seems to be that the homicide in the restaurant does not qualify as a war crime. The American officer is not acting in his role as a soldier, under the color of war. The dispute is purely personal.[24]

To summarize the results of our two hypothetical cases: the first stands for the application of domestic law and the implication that the civilian may not invoke the collective right of self-defense recognized under the law of war; the second stands for the application of domestic law instead of the law of war crimes. In both cases, the individual acts on his own, not as an agent of the collective, and therefore the only applicable law is the domestic law regulating individual behavior.

These examples of individual action invite us to consider more precisely what we are denying when we extricate the Polish farmer and the American officer from the collective represented by their respective nations and armies. If they are individuals, then what does it mean for the nation or the army to act as a collective unit? To answer this question, we need to reflect on the genesis and structure of the Rome Statute.

3.   COLLECTIVE CRIMES

The substantive crimes covered by the Rome Statute derive from three basic sources. The first is the Hague Conventions of 1899 and 1907, the second the Genocide Convention of 1948, and the third the Geneva Conventions of 1949, in particular, the Fourth Convention on the protection of civilians against the excesses of armed conflict. Of these three only the Genocide Convention even purports to impose direct criminal liability on the perpetrators of, and accomplices in, the five specified forms of genocide. The Hague and Geneva Conventions instead hold the contracting states accountable for breaches of the rules governing warfare. The Hague Convention imposes a duty on the responsible states to makes compensation to the states of the victims; the Geneva Conventions go further and impose on each High Contracting Party the duty to enact penal legislation to punish grave breaches of the Convention,[25] the details of which will concern us now.

A.   *Aggression and War Crimes*

In 1996 the United States enacted legislation to punish such grave breaches, but limited the jurisdiction of U.S. courts to crimes committed by or against American nationals or by or against members of the U.S. military forces.[26] In 2002, Congress changed the name of the offense to "war crimes," for the first time introducing liability for war crimes into the U.S. system of legislation. All eight of the grave breaches, together with specific prohibitions in the Hague Conventions, are carried forward in Article 8 of the Rome Statute. Yet something curious occurred in the nature of these offenses when they became the basis for individual liability at the International Criminal Court: they were subtly transformed into collective offenses. The concept of the war crime has undergone a transformation that is precisely the opposite of what is commonly believed.

The Hague Convention prohibits soldiers from engaging in specific forms of warfare, such as "employing poison," killing or wounding "treacherously," and employing "arms, projectiles, or material calculated to cause unnecessary suffering."[27] All of these offenses might be committed by a single soldier acting alone. Of course, under the Hague Convention, the soldier is not personally liable for a war crime. Indeed, the terms "crime" and "breach" are not used in the Convention; rather, "a belligerent party...shall be responsible for all acts committed by persons forming part of its armed forces."[28] The remedy against a responsible belligerent is a claim for compensation.[29] According to Article 3, liability attaches to the state of which the soldier is a national. In line with the traditions of international law, states are liable only to states. The state of nationality takes responsibility for a wayward soldier's crime.

Building on Nuremberg and other post–World War II military trials, the Rome Statute purports to impose liability directly on the "natural persons" who commit war crimes. This is the genesis of international criminal law, which trains its gaze on individual punishment, as opposed to international law proper and its focus on the relations between nation-states. But paradoxically, the transition from the Hague Conventions (1907) to the Rome Statute (1998) has rendered the applicable crimes collective in nature. This is a little known fact, but an elementary review of the law establishes the point.

Take a simple case of employing poison in warfare in violation of Hague Convention IV Article 23(a). Any individual soldier who uses poison in warfare violates the Convention. Today, a violation of the Rome Statute requires much more. Article 8 begins ambiguously by referring to all war crimes and then limits the concern of the International Criminal Court to those cases "in particular when committed as a part of a plan or policy or as part of a large-scale commission of such crimes." Because no treaty has defined the term "war crimes,"[30] we should look to the Rome Statute itself for a definition. But the Statute waffles and informs us only that prosecution will take place "in particular" when the violation is carried out in a way that reflects the collective policy of the army, the division, or the battalion.

This might be a good example of what the once fashionable French philosopher Jacques Derrida would call "deconstruction": the text of the Rome Statute strives for individual responsibility, then undermines itself by stating the opposite of what it purports to hold. The drafters would like to impose individual liability, but they all come from a tradition in which collectives—armies, and indeed nations—confront each other on the field of battle. Thus they extend one hand toward individual responsibility and with the other confirm the collective nature of warfare. When the federal criminal code punishes war crimes in 18 U.S.C. § 2441, we cannot be sure whether the relevant conception of the war crime is that of a single individual acting alone or, as suggested in the Rome Statute, the collective offense of a crime committed "as a part of a plan or policy or as part of a large-scale commission of such crimes."

This ambivalence about the nature of war crimes runs more deeply in the law of war than one might imagine. In both the Hague and Geneva Conventions—precursors to the Rome Statute—the drafters cannot decide whether war crimes are the acts of single soldiers or of units of soldiers acting in cooperation. Reflect for a moment about the war crimes that are called grave breaches, listed in Article 147 of the Fourth Geneva Convention:

1. wilful killing,
2. torture or inhuman treatment, including biological experiments,
3. wilfully causing great suffering or serious injury to body or health,
4. unlawful deportation or transfer or unlawful confinement of a protected person,
5. compelling a protected person to serve in the forces of a hostile Power, or
6. wilfully depriving a protected person of the rights of fair and regular trial prescribed in the present Convention,
7. taking of hostages, and
8. extensive destruction and appropriation of property, not justified by military necessity and carried out unlawfully and wantonly.

All of these offenses are repeated, in almost precisely the same language, in Article 8(2)(a) of the Rome Statute. But is it coherent to treat these violations of the Geneva Conventions as crimes committed by individuals acting alone? We can do this, if we so choose, with regard to the first three grave breaches: willful killing, torture, and causing great suffering. But what about deportations? How does a single soldier, acting alone, deport an enemy soldier? Deportation is not simply chasing someone into another country. It requires a legal process, an entire apparatus of judgment and enforcement. One individual "deporting" another is a practical impossibility.

The same is true about all the other grave breaches except the last, the unnecessary destruction of property, which an individual soldier can certainly do by himself or herself. But try to imagine an individual embedded with a unit compelling someone to serve in the unit, or depriving him of a fair trial, or taking a hostage all by himself.

These breaches are collective for two reasons. It might be the case that the breach is possible only with the complicity of military or governmental officials—for example, depriving someone of a fair and regular trial. In other cases, the individual cannot act alone because the event is public and other members of the unit must either endorse or reject the wayward action—for example, taking hostages. One way to resolve this tension is to approach these crimes with the understanding that although some may occur on the front lines, on the sole basis of individual initiative, others occur only when commanded and organized from above.

The other crimes of concern to the proposed International Criminal Court are all connected in one way or another with the concept of war. Let us think first about the crime of aggression. Though the United Nations still cannot agree on a formal definition, the concept of aggression serves a critical function in the entire structure of war and self-defense. The concept of aggressive war triggers the right of the attacked state to exercise collective self-defense.[31] There are problems in defining the threshold of intrusion that should constitute aggression, as there remains a dispute about the relevance of "just cause" to designating aggression a crime.[32] But despite these uncertainties, there is no doubt about the collective nature of the offense; the paradigm is one state's army marching across the territory of another. If a single Arab American tourist throws rocks at Israeli military installations, that is an act of vandalism, not an act of aggression. The group dimension of aggression carries with it the suggestion of recurrent and committed battle with aims of conquest, or at least of settling some political dispute.

The word aggression is normatively loaded. It implies that there is no adequate justification for the use of force. In this respect it resembles terrorism, about which there can also be much debate regarding violence "for a good cause." In a recent British decision, the House of Lords concluded that aggression was illegal under customary international law and overturned earlier British decisions that argued that the crime's contours were so contested that it could not be considered an international offense.[33] The House of Lords declined to provide an exact formulation for what counts as aggression, preferring instead to rely on the historical precedent of Nuremberg. Although the charge of conspiracy to commit aggressive war was controversial in 1945, it nonetheless resulted in a prosecution.[34] Consequently, if the crime was sufficiently defined in 1945 to allow for an international prosecution, it would be incredible to insist that the crime had somehow become *less defined* during the intervening years. This argument liberated the Lords from providing a full definition of the crime, but the task of providing a definition still remains for the international community. However the definition is finally resolved, aggression will serve as a conclusory label identifying a collective crime against the international order.

For the purposes of criminal liability, however, a clear definition of collective aggression will hardly solve the problem of individual liability. Assuming that the German invasion of Poland would satisfy any reasonable definition of aggression, it could not have been the case that every German soldier who participated in the invasion was guilty of a criminal offense. The hurdle is not only the practical difficulty of prosecuting millions; the deeper problem is that wars are not fought by collections of individuals who might be held liable for their conduct. Wars are fought by armies trained to fight under a unified command. Their effectiveness depends on inculcated discipline and sanctions imposed for disobedience. The collective army might be liable for aggression, but nothing about the guilt of individuals follows from this charge.

The tension between collective and individual responsibility poses a number of serious historical questions. Why did the Geneva Conventions use such obviously collective categories, such as "wilfully depriving a protected person of the rights of fair and regular trial"?[35] No single person can do this alone; if there is a deprivation of a fair trial, there might be hundreds or thousands collaborating in what they believe is a legitimate practice (e.g., depriving the detainees in Guantánamo Bay of a hearing). The Rome Statute carries forth this language without recognizing the problem of specifying who can be liable for committing the offense. The Hague and Geneva Conventions could use formulae of this sort because their purpose was not to impose individual criminal liability but to hold states accountable for not having prevented the collective activity.

It is important to remember that the charge levied against the general staff in Nuremberg was not aggression, but conspiracy to engage in aggressive war as a crime against the peace. The common law concept of conspiracy captured the collective dimension of the German leadership. There can, of course, be a conspiracy to wage war without an ensuing war, but in fact the concept of collective leadership used in Nuremberg was applied after the application of an effort to identify those responsible.

As there are problems with holding every soldier liable in a collective act of aggression, the same quandaries arise in many war crimes. Take, for example, the newly drafted war crime of an occupying power "transfer[ring] directly or indirectly...parts of its own civilian population into the territory it occupies."[36] This provision amended and broadened a provision of the Geneva Convention that applied only to direct transfer of a population. The obvious political purpose of the amendment at the Rome Convention in 1998 was to stigmatize the Israeli settlements of the West Bank. Difficult issues confound the question of whether the subsidized settlements of the West Bank constitute transfer, which one would think was supposed to describe a forcible relocation of persons.[37] There seems to be no basis, in legal practice or legal theory, to hold the settlers themselves liable for a war crime.[38]

To review these points about aggression and war crimes, the issues of individual and collective responsibility are confounded by two factors. First, the historical transition from the Hague and Geneva Conventions to the Rome Statute has, *malgré tout*, intensified the collective dimension of war crimes. Second, many of the terms of the Hague and Geneva Conventions were (and are) collective in nature, whether they are discussed that way or not. Both of these factors elicit an element of international criminal responsibility that has yet to receive appropriate notice and analysis.

The practical implication of this argument is that it casts radical uncertainty on criminal liability for war crimes under 18 U.S.C. § 2441. When offenses are collective in nature but only individuals are prosecuted, the field is wide open to prosecutorial discretion. This is obviously not a sound situation when seeking to develop a law of war crimes that conforms to the principles of legality and nullum crimen sine lege (no punishment without prior legislation).[39]

B.   *Crimes against Humanity and Genocide*

When we turn from aggression and war crimes to the third category—crimes against humanity—the collective nature of the required action is again apparent on the face of the statute. The specified crimes, which include murder and rape, must be "part of a widespread or systematic attack directed against any civilian population."[40] Of course, it might be technically possible for a single individual, without the aid of an organization, to carry out a "widespread or systematic attack." Perhaps the Unabomber, Ted Kaczynski, came close to meeting this standard. But the Rome

Statute makes it clear that individuals acting alone cannot commit crimes against humanity. The "widespread or systematic attack" must be "pursuant to or in furtherance of a State or organizational policy to commit such attack."[41] The words could not be clearer. In our view, the phrase "organizational policy" should inform the interpretation of war crimes.

In light of the history of crimes against humanity, this was to be expected. The origin of the concept lies in states' systematic abuse of their own citizens. When the Nazis committed atrocities against Polish Jews, they committed an international offense against Poland. When they tortured and murdered their own people, they committed no offense then recognized by international law.[42] Thus the Nuremberg tribunal had to recognize a "crime against humanity" in order to prosecute the Nazi leaders guilty of murdering German Jews. From the very beginning, therefore, the aim was to censure the way dominant cultures sometimes persecute minorities in their midst.

The fourth category, genocide, is a spin-off from crimes against humanity. Nuremberg was, of course, in large part about genocide, but the concept had yet to become entrenched in international legal discourse.[43] The Genocide Convention of 1948 represented a major step in the development of individual criminal responsibility, and it was a stimulus to claims of universal jurisdiction as well.[44] The structure of the Genocide Convention is carried forward in the Rome Statute, Article 6. More than any other statutory formulation, Article 6 appears to be aimed, at least nominally, at individuals. Genocide is committed by those who engage in one of five specified actions against groups with the prohibited purpose, defined as the "intent to destroy, in whole or in part, a national, ethnical, racial, or religious group, as such." It looks as if a single individual, acting alone, could have this intention. Suppose a Sinophobe is walking down the street in New York. He kills the first two Chinese people he sees, with the intention of destroying the Chinese people, at least in part. Technically, he has committed genocide. Did the drafters of the Genocide Convention really mean to target this kind of case?

Our shared understanding of genocide, based on the historical paradigm of Auschwitz, derives from deep-seated group hostility. The essence of the Holocaust was one nation seeking to eliminate another. Some individuals were especially guilty and therefore properly subject to prosecution for murder, but the collective "eliminationist" background informed and directed their actions.[45] In the final analysis, our intuitive understanding of genocide should help guide our interpretation of the statutory definition.

This shared intuitive understanding of genocide both expands and restricts the literal definition. According to the terms of the Statute, the prohibited intention must be directed toward a "national, ethnical, racial, or religious group." But if one large group of people seeks to eliminate another, the words "national, ethnical, racial or religious" may fall short of our intuitive understanding of genocide on the ground. In the case of bloody conflict between the Hutus and the Tutsis, for example, the differences are arguably not national, racial, or religious, and it is not clear whether the word "ethnical" captures their historical socioeconomic differences.[46] Yet neither the United Nations nor the international community at large has had qualms about applying the crime of genocide to the Hutus' persecution of their rival group. The reason that the international community can respond so clearly to collective persecution in Rwanda is that the motivating force behind the law is not the letter of the 1948 treaty defining genocide, but a historical paradigm of killing in order to eliminate a *genos* from the human species.

Our point has been made. Despite the ambitions of the Rome Statute to impose liability on natural persons, the drafters in fact had in mind collectives meeting on the field of battle. They conceived of collectives engaged in aggression, in the "plans or policies" necessary for war crimes, in the "widespread or systematic" abuse characteristic of crimes against humanity, and in the collective hatreds that stimulate crimes of genocide. Perhaps the drafters were simply too wedded to the idea of collective confrontation in warfare to conceive of totally individualistic crimes at the level of international criminal law. Whatever the reason, we have a body of law that clearly deconstructs itself. At one level it purports to address individual legal accountability. At every other level, though, it accentuates the role of collective action in bringing about the crimes of "concern to the international community."

4. THE RELEVANCE OF GUILT

If we know that collectives confront each other in warfare and in committing the crimes within the jurisdiction of the International Criminal Court, we need to ponder another elusive question: Which collective enters the field of battle? Who is present when an individual is killing civilians or torturing prisoners? The traditional answer to the first question is that the entire nation goes to war. Lieber understood this clearly when he stressed that nations must "enjoy, and suffer, advance and retrograde together, in peace and in war."[47] This collectivist understanding of the nation at war expressed the Romantic urgings of early nineteenth-century thinkers who sought the fulfillment of their personal and national quest for glory by seeking out the field of military honor.[48]

The Geneva Conventions took a major step in separating the fighting forces from the nation they represented. The idea that civilians were "protected persons" implied that they were not at war, and they should not be vulnerable to attack in the same way as the military. Some observers go so far as to suggest that in the war between Germany and the United States, the U.S. army was bound to protect the lives of German civilians as much as they were to pay heed to the lives of Americans on the sidelines of battle.[49] But once we say that only the army and the government constitute the collective at war, then we can begin to subdivide even further. This is why we, as a nation, were so confused about who was responsible for the torture in Abu Ghraib. Was it just the torturer's unit, the army in general, all military personnel, the Defense Department, the government as a whole, or indeed the entire American population who were responsible for having allowed the military to escape scrutiny? Like an accordion, the field begins with an individual and expands to include the entire nation. How do we know where to stop? How do we define the relevant collective?

The International Court of Justice delved into this matter in 2007 when Bosnia argued that Serbia—as a state—should be held legally responsible for the genocide against Bosnian civilians at Srebrenica.[50] Although this might seem obvious, the ICJ case was the first time that the World Court considered state responsibility for genocide, as opposed to mere criminal liability for individuals involved in planning the genocide. The ICJ concluded that any acts performed by a "state organ" are attributable to the state as a whole, according to basic principles of state responsibility under customary international law.[51] As a specific application of this general principle, acts of genocide by a state organ could be attributed to an entire state and yield collective responsibility under international law. The result of such collective liability would include a responsibility to make financial reparations for the damage.

Although the ICJ recognized the basic principle of collective responsibility for genocide, it declined to hold Serbia directly responsible for the genocide at Srebrenica. The court concluded that under the Genocide Convention Serbia was responsible only for failing to prevent the genocide and for failing to cooperate with the International Criminal Tribunal for the Former Yugoslavia for refusing to hand over key suspects, including Ratko Mladić.[52] In an incredible act of myopic legal reasoning, the court held that General Mladić's actions could not be attributed to the Serbian army, according to the internal law of Serbia, or to any other official organ of the Serbian government.[53] The court concluded that Mladić's actions could be attributed only to Republika Srpska, despite the fact that the general may indeed have been "administered" from Belgrade and even received a promotion to colonel general from Belgrade.[54]

In a blistering dissent, the court's vice president noted that Serbia's involvement in the genocide was supported "by massive and compelling evidence."[55] He also worried that the "inherent danger" of the majority's approach was that it would give states "the opportunity to carry out criminal policies through non-state actors or surrogates without incurring direct responsibility." Finally, the dissent also pointed—with incredulity—to the majority's argument that genocidal intent could not be attributed to the Serbian people because their desire for a Greater Serbia, in the words of the majority, "did not necessarily require the destruction of the Bosnian Muslims and other communities, but their expulsion." One might have said the same thing about the German desire for a Jewish-free Europe.

The general point of the dissent was that you cannot attribute actions to the state in exactly the same way that you attribute actions to individuals at a regular criminal trial. One ought to be more sensitive to general patterns of conduct. Although this is anathema in a regular criminal trial, where the conduct of individuals is at stake, the dissent is correct to point out that it seems much more appropriate where the culpability of an entire state is at issue. The court failed to draw even the most basic inferences from the facts on the ground in Bosnia. Perhaps these factual connections would have been easier for the ICJ to accept if Slobodan Milosevic had not died before his trial ended at the tribunal. His premature death robbed the world of a binding judgment against him for genocide. Nonetheless, there was sufficient evidence presented at Milosevic's trial, and the ICJ should not have needed a final judgment in order to accept that evidence.[56] It is hard to deny that there is sufficient connection to a state organ when the individual in question is the head of state.

Even assuming that the ICJ was correct in concluding that Serbia should escape legal responsibility at The Hague for the Srebrenica massacre, this would not end our analysis. Surely, there is some greater responsibility here, even in cases where genocide cannot be attributed to the *state* through the customary rules on state responsibility. We have an intuitive sense that a greater collective responsibility is at work, although locating and identifying that sentiment is rather difficult.

Our best guide is an assessment of collective guilt. Surely more than the individual perpetrators are guilty, but it is not easy to construct the relevant community or collective that must stand up and claim responsibility for wrongful actions. Perhaps we can further our analysis by considering another example of collective responsibility for individual wrongs, one closer to home and requiring greater introspection. Some might say that in cases like Abu Ghraib the relevant question is not guilt, but shame. Maybe both sentiments are appropriate.

Seeing the photographs of humiliation in Abu Ghraib, many of us felt acute shame in being part of a nation that could go to war with righteous ideas and end up replicating, if not aggravating, the abuses of the rogue state we called our enemy. Shame is a sentiment of the heart, often a function of irrational reactions to the gaze of others. Those with physical deformities may feel shame about their body. Even parents and children can, irrationally, feel shame about the behavior and sometimes the mere presence of the other.

Guilt is a more rational sentiment, based more on what we do than who we are. Neither the vast majority of U.S. soldiers nor Americans as individuals have done anything wrong in Iraq (apart from the invasion itself), and thus might balk at allegations of collective guilt for the atrocities. Yet in other cases of collective action, we willingly affirm collective guilt and a shared duty to make reparations. This was the widely accepted approach toward German liability for the Holocaust, and there are many who urge the same approach toward America's responsibility for slavery.

One way to think about guilt versus shame is to begin with the response that fits our sentiment of responsibility. Guilt represents a debt. The proper response to such a debt is to suffer punishment or to pay reparations to the victims. Shame invites a retreat from the public eye. If you are ashamed, you do not expose yourself to punishment, nor do you extend your hand in a gesture of repair. When you are ashamed, you cannot bear the critical gaze of others: you hang your head low. Though former Secretary of Defense Donald Rumsfeld proposed compensation to the victims of abuse at U.S. military hands, it is hard to see this offer as expressing either guilt or shame; it seemed more like an effort to buy silence. If compensation were coupled with a finding of high-level American wrongdoing, we would get closer to an act of atonement.

If we are trying to determine the collective that is expressed in the actions of going to war and committing crimes in the name of that war, then a shared sense of guilt might be the most rational guide to an assessment of those collective actions. If the nation feels guilt for its actions, this is a clear sign of a collective consciousness that it was the nation that had gone to war and had also engaged in the atrocities for which it is blamed. If only the army or a political party feels this guilt, then these sentiments point to a different collective consciousness. This may be too solipsistic for some, but it does capture the idea that when a collective feels shame—or its opposite, pride—it has taken ownership of the actions in question.

The dark side of collective guilt, particularly as perceived by others, is that the attribution of guilt often provides grounds for collective punishment. This is particularly evident in Middle Eastern politics today. Indeed, in the Israeli-Palestinian conflict, collective action and collective punishment seems to be all there is. In the heat of that conflict, it is difficult to be seen as an individual, as someone who is simply doing his or her thing not as an Arab or a Jew, as a Muslim, Christian, or ultra-Orthodox Jew, but simply as a solitary man or woman who happens to live in this part of the world. There are many parts of the world that reveal a similar group consciousness—India and Pakistan, Northern Ireland—and they stand in sharp contrast to the liberal idea that the only true units of action in the world are individuals, not groups. In the Middle East, it is difficult to kill a member of the "other side" simply on grounds of personal hatred. Every killing implicitly invokes a confrontation between Palestinians and Jews, or between Islamists and settlers, or between terrorists and civilians.

This phenomenon makes the Middle East a fitting arena for reflecting on the issues of collective guilt and collective punishment. When a suicide bomber attacks Israeli children, Jews consider the entire Palestinian population guilty, directly or indirectly. When Jews move into the West Bank, establishing new settlements, Palestinians accuse the entire Jewish nation of taking Palestinian land and creating facts on the ground that impede the creation of a Palestinian state. This reciprocal perception of the other side's collective guilt fuels the endless cycle of violence that has tragically dispelled dreams of peace in the region. The two nations, Jews and Palestinians, are reduced metaphorically to single agents struggling against each other.

Beginning in September 2000 under the pressure of the second Intifada, many otherwise sober and rational American Jews lost their sense of proportion and began advocating collective punishment of Palestinians. On the assumption that it takes a village—literally—to create a suicide bomber, Alan Dershowitz started proposing various ways of penalizing the entire village, including destruction of all the houses in the area. In his own words:

Israel's first step in implementing this policy would be to
completely stop all retaliation for five days. Then it would publicly

declare precisely how it would respond in the event of another terrorist attack, such as destroying empty houses in a village used as a base for terrorists, and naming the village in advance. The next time the terrorists attack, the village's residents would be given 24 hours to leave, and then the Israeli troops would bulldoze the houses.[57]

Dershowitz's colleague, the Washington lawyer Nathan Lewin, went further and proposed the death penalty for the entire family of the suicide bomber. Lewin brazenly invoked the precedent of Amalek as a biblical warrant for collective punishment.[58] Both of these sophisticated, liberally trained lawyers sense that their proposals seem outrageous to others; therefore, both guard their flanks with some traditional legal arguments of individual responsibility. Dershowitz shifts subtly from collective punishment to punishment for causal complicity in the acts of terrorists. There is no doubt that particular individuals who aid and abet the commission of terrorist acts—by providing money or weapons, counseling or encouragement—should be guilty as accessories to the murder. As to the handlers, who convince young men and women that by killing innocent children they acquire a place in Heaven, the maximum punishment would be appropriate. But it is not clear how far we should stretch the idea of complicity to sweep up the "good" Palestinians who implicitly endorse suicide bombings but take no active steps to facilitate such attacks, or to family members who do nothing but share a bloodline.

To buttress his argument for Palestinian complicity, Dershowitz relies heavily on a case in which men in a bar who cheered on a group of rapists are presented as potentially complicitous in the rape itself.[59] Dershowitz says that in such a case there is nothing wrong with thinning out the criteria of complicity so long as "the consequences imposed on the [accessories] are proportional to their complicity."[60] That is, if you actually raped the victim, you might get twenty years; if you held her down, ten years; and if you merely cheered on the primary offenders, you might get a year in jail. This is absolutely correct, even though the U.S. legal system and many others hold that all aiders and abettors may be punished to the same degree as the actual perpetrators.[61]

Bear in mind that the law of complicity is firmly grounded in liberal criteria of individualized justice; it has nothing to do with collective punishment. The latter only comes into play when an entire village might suffer the destruction of its houses, whether or not the residents were directly complicitous in a particular act of terror. And even if they are theoretically supportive of a suicide bomber in their midst, Dershowitz hardly envisions a time-consuming, individualized trial of each person who might lose his or her home under the plan. The leap from the collective guilt of the village to its collective punishment would be automatic.

To alleviate the obvious injustice of punishing innocent individuals as well as the guilty, Dershowitz claims that the prior warning generates a new basis for blaming those who suffer:

> The policy and its implications will be perfectly clear to all the
> Palestinian people: whenever terrorists blow themselves up and kill
> Israeli citizens, they also blow up a house in one of their villages.
> The destruction is entirely their own fault, and it is entirely
> preventable by them.[62]

Note the moves made in this argument. First, the intervening agency of the Israeli army drops out of the picture. The image is one of self-destruction. The criminal becomes the agent of his own punishment. But here the "criminal" consists of many different people who may or may not be the same as those who suffer the destruction of their homes. Dershowitz's argument becomes plausible—if it *is* plausible—by collapsing an entire nation into a single actor and then eliminating the Israeli army as the intervening agent. In the rhetoric of self-destruction, "terrorists" choose the punishment of their fellow nationals. Thus Dershowitz tries to justify indiscriminate devastation by an advance warning.

The law of war has taken pains to guard itself against these sorts of maneuvers. If people heed these warnings, the attacking military would become a cheap way of terrorizing and destabilizing the population. They would repeatedly warn and rarely attack. Ultimately, the argument for collective punishment in this scheme is not justice but deterrence.[63] The same is true of Lewin's rage-filled fantasies of hanging entire Palestinian families. From the Kantian perspective, wherein all rational beings must be treated as ends in themselves, deterrence is a violation of human dignity because it represents the use of a condemned person merely as a means to achieve the end of social protection.[64] (When individuals are punished to deter future crime, the punishment has little to do with the intrinsic nature of the criminal or his crimes, but more to do with how we might benefit society by making him an example; this is precisely the style of reasoning forbidden by Kant's notion of the Kingdom of

Ends.) Furthermore, there is no evidence that the violent reprisals on the West Bank have had much of a deterrent effect on terrorists and suicide bombers. On the contrary, punishment perceived to be unjust has the effect of increasing the solidarity and resentment of those who suffer and, ultimately, the effect of augmenting resistance rather than decreasing it.

But neither the argument of complicity nor the shift to deterrence can assuage our sense of injustice about blaming and punishing the collective for crimes actually carried out by individuals. This leads us to suspect that behind these rationalizations for collective punishment lurk deeply held sentiments of collective guilt. The proponents of collective punishment assume that Palestinians are guilty as a collective for nurturing a culture that takes pride in suicide bombers. This is not an unreasonable assessment of the way the entire culture contributes to the actions of a few.

To round out the picture, however, we should consider the way Palestinians attribute collective guilt to Jews and Israelis, and how they then justify the collective punishment of all Jews as a proper response. Hamas and other right-wing Palestinian terrorist groups regard all Israelis, and probably all Jews, as guilty of the great sin of settling the country and defeating the combined Arab armies in the War of Independence. The continued presence of the Jews on Palestinian ancestral land recreates the crime in every generation and seems to represent an ongoing humiliation to Arab honor. It is not surprising that the Muslim world has become a fertile market for all lies manufactured against the Jews, from early Christian myths to the latter-day *Protocols of the Elders of Zion*. Israel's partnership with the United States only exacerbates the image of Jews as exercisers of uncanny powers, able to conquer and manipulate the world's media and financial systems.

The idea that Jews act as a corporate body obviously has its origins in the Book of Matthew, where the Roman governor Pontius Pilate decides to deliver Jesus to his death, and yet as a Roman and as an individual, he is able to wash his hands of guilt and attribute the entire decision to the Jewish crowd: "And all the people answered and said, 'His blood be on us and on our children'" (Matthew 27:25). However large this crowd might have been, it surely did not include all the Jews then living, not to mention the Jews of future generations, and yet all the Jews then living and not yet born are implicitly held accountable as a corporate entity for the behavior of this crowd outside the palace of Pontius Pilate. The words of the crowd, "and on our children," are fashioned to entail liability for future generations.

Holding the Jews liable as a corporate body is no different from the way the Jews hold Amalek guilty across the generations for some mythical crime committed against Moses in the desert—attacking from the rear, according to Deuteronomy 25:18. Fortunately, we no longer know who the members of the tribe of Amalek are. If we did, those who take the commandments of the Bible seriously would face an embarrassing moral problem as to whether they were under a duty to continue to wage war against Amalek.[65]

The example of Amalek illustrates an important difference between collective guilt and a declaration of perpetual war. In the case of the Jews, the statement "His blood be on us" seems to be a self-attribution of guilt. There is no declaration of war against the Jews in Matthew, merely an ideological foundation for holding Jews forever guilty, both collectively and individually, for Christ-killing. But the notion of guilt is not used in connection with Amalek.[66] This is a case of a war declared in perpetuity, never subject to a peace agreement.[67]

The major puzzle behind these attributions of collective guilt is the scope of the group regarded as guilty. Even if the Jews in front of the house of Pontius Pilate had said "His blood be on us," what are the implications for the guilt of all Jews then living in Jerusalem, all Jews living then in the world, or indeed all Jews for generations thereafter? Even if there was some collective guilt deriving from the actions of the Jews, how far should the net be cast? This is no easy question, but it is indeed the same question we encounter in trying to assess collective guilt for war crimes in modern warfare. Who is guilty for the abuses of Abu Ghraib? Only those who actually administered the torture, the superiors who tolerated the development of their culture of abuse, the entire U.S. army in Iraq? Or does the guilt spread even further and taint all Americans for having supported, or at least tolerated, the ill-conceived war that generated these excesses of sadism?

To make some progress on this issue we must consider the development of collective guilt in Western thought. The Bible is a suggestive place to begin.

5. GUILT AS COLLECTIVE POLLUTION

Interestingly, the biblical terms for crime and punishment are closely related. Thus a debate persists about what Cain was actually saying when he appealed to God, "My *avon* is more than I can bear" (Genesis 4:13). The same word *avon* can be translated as either "iniquity" or "punishment."[68]

Our entire understanding of Cain as a person depends on this word. The translation "punishment" leads to the traditional view that Cain grovels before God to save his life. He cannot own up to the evil of this deed. The alternative translation dignifies Cain by having him recognize the enormity of his offense and confess before God that he cannot bear it; it comes close to an admission of guilt.

As it is first used in Genesis, the concept of guilt, or *asham*, carries the same multiplicity of functions. It is interwoven in a story told, intriguingly, three times, as if one telling would not make the point. The pattern is always the same: one of the forefathers of the Jewish people is about to enter a foreign land, where he suspects that the "barbarians" will kill him and take his wife. Therefore, Abraham (twice) and Isaac (once) revisit the same deception: each tells a foreign potentate that his wife is, in fact, his sister. In all three cases something happens to inform the potentate that either he or a man of his court is about to commit adultery.

In the first version, Abraham (then called Abram) passes off Sarah (then called Sarai) as his sister to the Egyptian pharaoh, who takes her into his court. Plagues then descend upon "Pharaoh and his household" as a sign that a sexual sin has occurred or is about to occur. The sin is disclosed very much as it is in Sophocles' play *Oedipus Rex*, by a manifestation of natural disorder.[69] Pharaoh quickly realizes that something is wrong in the natural order and confronts Abram with his lie. Note the nature and scope of the taint of adultery: it affects the entire household, but apparently not the entire village or county. The collective guilt is limited, it seems, to those who might come into contact with Sarai and be tempted to sleep with her.

In the later retelling of the same basic story (with Abram renamed Abraham and a potentate named Abimelech), the truth of sexual sin is realized not by a plague but by God coming to the king in a dream and saying, "You are to die because of the woman that you have taken, for she is a married woman" (Genesis 20:3). In this retelling it is not clear how many people are affected by the pollution, other than Abimelech himself. In the third telling, when Abraham's son Isaac passes off his wife, Rebecca, as his sister, a king (also named Abimelech) discovers the lie when he sees them engaging in affectionate behavior that would be incest if they actually were brother and sister. Assuming that they are not incestuous siblings, Abimelech confronts Isaac, exposes the lie, and then says, "What have you done to us? One of the people might have lain with your wife, and you would have brought guilt [*asham*] upon us" (Genesis 26:10).

It is worth noting how the epistemology of discovery changes across the three stories. In the first, pharaoh concludes from the aberrance in nature that there must be a sexual abnormality in the kingdom. In the second story Abimelech learns of the potential sin because the voice of God comes to him in a dream. In the third story Abimelech uses his own eyes and common sense to discover that something is untoward. There is an increasing secularization and humanization of the process, and the same process is found in the treatment of guilt itself. At the beginning guilt is collective, objective, and a stain upon the land. It can be cleansed only by a guilt-offering in the temple (Leviticus 5:6, 15–16, 18–19, 25).

Yet the nature of guilt does not quite evolve in line with the epistemology of its discovery. In the first story, the guilt is understood as something like a stain on the household. In the second story, when God comes to Abimelech in a dream to warn him of the impending adultery, Abimelech responds that he is personally innocent in his heart (she told him that she was Abraham's sister) and pleads, therefore, not only for himself but for his entire "righteous nation" (Genesis 20:4). Similarly, in the third telling, Abimelech accuses Isaac of bringing guilt "upon us," on his own people—implicitly not just on Abimelech, and apparently not at all on Isaac or Rebecca (Genesis 26:10). The guilt arises not from the culpability or deed of the offender but solely from the consequence: the participation in adultery. And the guilt attaches collectively to the entire nation represented by the person who engages in sinful intercourse with a married woman.

This is how the notion of guilt makes its appearance on the biblical stage. In those places where you would expect to find it—after Adam and Eve eat the forbidden fruit, after Cain kills Abel, after Ham abuses his father, Noah—the concept is absent. Adam and Eve feel shame, and Cain complains that his sin (or punishment) is too great for him to bear. But none of these characters prior to Abraham mentions possible guilt for his or her misdeed.

The remedy for guilt, in the way the term is used in the Hebrew Bible, is to bring a sacrifice. The sacrifice cleanses the stain. Remarkably, the word used repeatedly in chapter 5 of Leviticus to describe a whole range of polluting events and the appropriate sacrificial response is also asham (Leviticus 5:6, 15–16, 18–19, 25). These sacrifices, called "guilt sacrifices," atone for various polluting events, such as touching an unclean animal (Leviticus 5:2). The prescription is to bring a guilt sacrifice to atone for guilt; the same word is used both for the cleansing act and the stain. The interplay here between the remedy and the deed recalls the controversy about translating the word that Cain uses in his complaint that something about his fratricide is too difficult to bear. Some think that he is referring to the punishment, others to the crime itself. This easy interchange of the negative and the positive, the contamination and the decontamination, reveals the tight conceptual connection between the two.

The hypothesis seems safe that the ancient world understood these concepts in a different way from our own understanding. In contemplating whether Oedipus feels guilt or shame for his fated patricide and incest, it is often said that the Greeks at the time of Sophocles did not distinguish between the two concepts.[70] There are signs of both in the play. When Oedipus discovers his crime, he craves punishment as though he were guilty in the modern sense, but the method of his self-inflicted punishment—putting out his eyes and going into exile—resonates with shame. He cannot bear to see others looking at him.

Although the ideas of guilt and shame are interwoven in Athens, they are distinct in Jerusalem. The biblical text recognizes a culture of shame in the story of Eden and a distinct understanding of guilt and guilt-sacrifices in Leviticus. Even in Athens, there are clear differences between Sophocles and Aristotle, who was born a century later than the playwright. The Nicomachean Ethics continues to be a guide to the general theory of responsibility, enabling us to understand the concept of guilt as it is used in the modern sense.[71]

The ancient sense of guilt as pollution is still with us. It is expressed in the conceptual connection between contamination and decontamination. The form that this connection takes today is the belief that a guilty act requires punishment, and the metaphors that we use to discuss retributive punishment carry forward the principle of decontamination. We share the Hegelian faith that punishing a wrongdoer vindicates the Right against the Wrong or validates the norm against those who would undermine it. In this way of thinking, the crime pollutes the moral order and the punishment serves to restore the law and the world as it should be.

In the modern approach to guilt, the focus is not on pollution but on the feelings of those who are guilty. The shift is from guilt's external impact on the world to the inner, human experience of guilt. Unexpectedly, the book of Genesis is a powerful source for this subjective understanding of guilt, and a careful reading of the Joseph story reveals that the subjective perception of guilt has a more ancient vintage than we might ordinarily think.[72]

The saga begins with a built-in conflict between Joseph and his ten elder brothers. Jacob, their father, loves Joseph the most, and when some brothers receive more love than others, as Abel was favored by God, one can always expect enmity between them. The conflict among the sons of Jacob becomes more acute when Joseph relates two dreams, which his brothers interpret as fantasies of domination over them.[73] The brothers conspire to kill Joseph and then throw him into a pit. Reuben protests the plan and suggests that they merely leave him to die. This they do and then sit down to break bread, as though they are celebrating Joseph's demise. At that point, Judah sees a caravan of Ishmaelites approaching and realizes that it might be better to sell Joseph to the voyagers rather than kill him. (Apparently, it does not occur to him that selling their brother into slavery is also a wrong that they would have to conceal from Jacob and others.) Before the brothers can realize Judah's plan, a band of Midianites passes by. One of the groups (the text is ambiguous on this point) lifts Joseph out of the pit and sells him to one of the passing caravans headed for Egypt. Reuben discovers that Joseph has been taken and tears his clothes in distress. To cover up their crime, the brothers then dip Joseph's coat—Jacob's gift of love—in the blood of a slaughtered goat and take it to Jacob as proof of Joseph's death. The traveling merchants sell Joseph into the service of an Egyptian named Potiphar.

This is the end of the passage recounting the tale of crime and betrayal. It is worth noting that no one except the dissenter Reuben acts as an individual in this story. It is a fraternal collective that throws Joseph into the pit and, later, lifts him out. The brothers function as a unit. Even when Reuben protests, he speaks in the first person plural.[74] The next segment of the saga traces Joseph's rise to political power in Egypt. When he meets his brothers again, at least a decade later, he is the "governor of the land." With a famine in Canaan, Jacob sends ten of the brothers, excluding the youngest, Benjamin, to find food in Egypt. When they encounter Joseph, the ten bow down to him without recognizing him, but Joseph both recognizes them and recalls his dream.

There then follows a conversation that leads to the brothers' recognition of their guilt for the way they conspired to kill and abandon their brother—one of the most remarkable interactions in the corpus of biblical literature (Genesis 42:6–21). Joseph stages both a conversation and a physical environment that leads his brothers to understand the moral dimension of facts they had long known. The first step in the interaction is Joseph's accusing the brothers of being spies. It is hard to know whether Joseph himself believes the charge to be true or whether he is testing his brothers.[75] Joseph himself is acting as the officer of a state; his accusation of spying is designed to find out whether the brothers are the same or whether they identify themselves as a family rather than a nation. The brothers defend themselves by claiming that they are "the sons of one man in Canaan."[76]

The problematic aspect of the brothers' response to the spying charge is the seemingly gratuitous addition to their claim to be all the sons of one father: "The youngest is now with his father, and one is absent." This admission gives Joseph the opportunity to stage a dramatic recreation of one brother being absent. First, he suggests that the brothers send someone from their group to fetch their brother Benjamin. This proposal is never pursued. Instead, after holding them in custody for three days, Joseph suggests that they leave one of their collective in Egypt and return, as a group of nine, to fetch Benjamin.

At this point the brothers are moved to confess.[77] Great moral insights rarely arise from easily identified factors, but we can point to factors that separately or in combination might have generated the brothers' realization that they had committed a great wrong: (1) their spending three days in confinement, which somehow brought home to them the experience of Joseph in the pit; (2) Joseph's playing on their incompleteness as a set of brothers, first by insisting that they bring Benjamin to Egypt, then suggesting that they send home to fetch him, and finally requiring that one be left behind while the others seek to complete their number; and (3) Joseph's reminding them of their moral conscience, stressing that he himself is a "man of God." Though we do not know why the breakthrough ultimately happened, the beauty of the text enables us to see that only collectively could the brothers come to this realization.

But what, precisely, do they feel guilty about? Not about the supposed death of Joseph, nor throwing him into the pit with the intention of either killing him or letting him die. According to their own words, their guilt attaches to having heard and ignored his cries of anguish. Thus the guilt is displaced from the pollution to the act causing the pollution and, finally, to the victim's pleas to avoid committing the act. This subtle relocation of the guilt could either be trivial or profound. The trivial version derives from the way the brothers use their declaration of guilt to explain their current misery: "We saw the anguish of his soul when he pleaded with us and we did not grasp it, and therefore our anguish has come over us" (Genesis 42:21). Thus they explain their anguish as a consequence of having ignored someone else's anguish. Emphasizing the second part of the passage converts their confession of guilt into a tactical mistake about controlling their personal fate.[78] They realize that they could have avoided their current predicament—the psychological affliction of guilt—if only they had acted. This transforms the sentiment of guilt into something closer to regret.

A more profound interpretation of this event requires that we import some basic concepts from the philosophical literature on freedom of the will.[79] By analogy to the idea of second-order volitions as the mechanism for regulating and resisting first-order impulses, we should think of guilt as a second-order failure to resist our baser impulses. It is understandable that the brothers would want to kill one of their own who sought to rule over them, but they should have resisted their base homicidal impulses. Their second-order volition should have been to heed Joseph's appeal for compassion. It does not matter much whether that appeal is implicit in Joseph's humanity or is articulated as cries for help. The point is that the brothers did not hear it.

The metaphor of hearing correlates with actions producing guilt. We hear the voice of conscience rather than read an image of conscience in our mind, and thus it makes sense for the brothers to associate hearing with understanding the moral dimension of their actions. Further, Jewish theology emphasizes hearing over sight in the relationship with God. This is evident in Moses' confrontation with God on Mount Sinai and in the liturgical demand on Israel to hear and understand that God is one. By contrast, Christianity emphasizes the sense of sight and the role of images, particularly of Jesus on the cross, in sustaining faith. Significantly, the confession occurs as a collective event, as evidenced by the brothers speaking to each other before they declare their collective guilt. The impact of their confession appears to dispense with the need for punishment, for Joseph hears them and he cries (Genesis 42:24). The confession itself prepares the group for the reconciliation that occurs several chapters later.

The Joseph story has much to teach us about the theory of collective guilt. Even though they experience guilt subjectively, the brothers remain a well-defined collective. They confess to each other with a clear sense of inclusion and exclusion. This sense of well-defined contours eluded us in our earlier investigations of guilt on the national level. The problem is whether there is any way to express a national consciousness of guilt comparable to the brothers' awareness that they constitute a collective that has engaged in wrongdoing and that, as Lieber would have put it, they must "enjoy, and suffer, advance and retrograde together, in peace and in war."[80] Yet our inquiry into collective pollution and guilt in the biblical context does not resolve our major concern, which is to delineate the circle of collective guilt in cases of war crimes, torture, and other major atrocities. Where shall we look to get a better feel for collective guilt in modern warfare? Perhaps the relatively modern subjective sentiments of guilt and shame might be a better guide to the collective that stands together and takes the blame for wartime atrocities.

6. REVIVING THE ROMANTIC SOLUTION

In our ongoing quest for the correct unit for positing collective guilt in warfare, we are torn among the possibilities of locating responsibility and guilt at the level of the individual soldier, the military unit, the army as a whole, the governing political party, or the nation that has gone to war. The Romantic tradition has a distinctive suggestion, one that is worth taking seriously because the other advocates of collective guilt are unable to concur on the relevant matrix of social organization. Perhaps we must recognize that, after all, it is the nation that goes to war.

As we argued in chapter 6 when we discussed the rights of nations, the nation is defined by its organic cultural bonds: its language, its history, and sometimes its religion. The critical fact about national life is that the cultural bonds of language and history create a way of life that is unique and irreplaceable. In international law today we take the claims of national existence more seriously than ever. The law of genocide is designed primarily to avoid the killing of entire cultural entities. The cultivation of self-determination since World War I has largely been based on the idea that each nation has an inherent right to govern itself.[81] In chapter 6 we used this evidence, among other facts, to show that nations have a right of legitimate defense. But these ascriptions are a two-way street: if the nation has the right to live and to govern itself, then it arguably has the duty to stand collectively responsible for crimes committed in its name.

These ideas about the ontology of nations come more easily to some cultures than to others. Readers of the Bible have no difficulty recognizing the primacy of tribal nations in drawing the lines of identity and loyalty. Western European history was dominated by national identities for several hundred years. But the relevance of being Spanish, French, or German is now threatened from above and below—from above by the identity claims of being European, and from below by the local communitarianism of the Gael, the Breton, and the Corsican, to name a few. The conflicts from above are replicated in the Muslim world, where nation-states seek to establish their primacy relative to the claims of pan-Arab and pan-Islamic solidarity over questions such as the U.S. invasion of Iraq and the future of the Palestinian people. The struggle from below is played out in

Africa, where nations of Nigerians and Kenyans struggle for coherence in competition with loyalties owed to the Ibu and the Kikuyu.

In many respects we live today in the afterglow of the Romantic fascination with the nation as the relevant unit of glory in the dramas of international conflict. Of course, the Romantic movement of the early nineteenth century took many different forms. For most English poets of the period, the new emphasis was on the self and the spirit as they connected with nature. The banner was Wordsworth's dictum that poetry is "the spontaneous overflow of powerful feelings."[82] Yet a decade later Wordsworth was celebrating the Convention of Cinta, which allied British and Spanish troops against Napoleon. This was an ecstatic moment, bestowing upon the English soldier a chance for heroic greatness.[83]

The German Romantic movement had a more specific agenda, for it saw itself as engaged in a dialectic with the universalist claims of the French Enlightenment. As Isaiah Berlin brilliantly argued, Johann Georg Hamann (1730–88) was a critical figure in the early German reaction against the French claim—very like the American faith today—that all human beings everywhere are essentially alike.[84] Though he was friends with his fellow resident of Königsberg, Immanuel Kant, Hamann rejected the special province of reason and urged instead an antirational emphasis on holistic religious experience. Though the organism was one, the world was cast in provinces, unbridgeable units of national experience.

In the course of the nineteenth century, the Romantic movement had an impact not only on German music and philosophy but also on the course of German law. Savigny argued against codification in his famous monograph of 1814 on the ground that, although the French thought they had found the universal approach to law in the *Code civil*, the German nation would have to evolve on its own and find its distinctive path among the legal cultures of the world.[85] These artistic, philosophical, and legal themes interwove in the late eighteenth and early nineteenth centuries to generate a movement—a family of related views—that emphasized feeling over rationality, particularity over universality, and the impulses of genius over the predictability of normal life. In this skein of antirational sentiments, both the individual and the collective triumphed. Feelings became critical, and feelings about the nation, in particular, became the lodestar of international politics.

Germany's central role in this history set the stage for its national tragedy and the postwar preoccupation with collective German guilt. In the wake of the Holocaust, German intellectuals had no place to hide. As some of the best reflections on nineteenth-century Romanticism were written in Germany, some of the best assessments of collective guilt for the Holocaust also found expression in German. Among the latter we should refer to Karl Jaspers's classic work, *The Question of German Guilt*.[86]

Jaspers wrote immediately after the war, when there was a widespread tendency to regard the Germans as collectively guilty for all aspects of the war and to impose collective punishment by limiting the future development of German society.[87] Surprisingly, Jaspers comes out in favor of the idea of collective guilt but opposed to German guilt as an instantiation of the idea. To reach this conclusion, he situates the problem in a larger framework of four kinds of guilt: criminal, moral, political, and metaphysical.

Criminal guilt is the most familiar embodiment of the concept. There seem to be close associations between crime and guilt, on the one hand, and between guilt and punishment, on the other.[88] The term for guilt across Western languages (*culpabilité, Schuld, vina, asham*) is used uniquely in criminal law. We are familiar with the common law institutions of guilty pleas and guilty verdicts. The notion of innocence as expressed in the presumption of innocence is closely tied to the same concept. It is not exactly on point to say that someone who does not commit a tort or does not breach a contract is innocent. The term is part of a particular complex of ideas, including guilt and punishment, framed only by the criminal law.

Moral guilt may coincide with criminal guilt, but it need not do so. As Jaspers uses the term, the realm of morality focuses our attention on the inner quality of the deed, but not in the way we are accustomed to think today. Jaspers would say that those who act under duress or personal necessity—namely, those who may be morally and legally excused in the conventional understanding of those terms—are still morally guilty if they could have avoided the act. Thus Dudley and Stephens, the sailors shipwrecked at sea who consumed a cabin boy to survive, made themselves morally guilty, even though many commentators today would say that they should have been excused under the law.[89] No one blames them for submitting to overwhelming pressure, but they could have exercised heroic capacities to abstain from cannibalism and risk death by starvation. Their failure to do so was enough for them to be morally guilty, though the Crown commuted the sentence to six months in prison.[90]

Both political and metaphysical guilt are beyond the moral category of the avoidable. They attach even in cases of living under dictatorships, where it is not humanly possible to avoid the inhuman actions of those in charge. Political guilt is borne by each person in a political community merely by virtue of being there and being governed. As Jaspers puts it in a disarming sentence: "Es ist jedes Menschen Mitverantwortung wie er regiert wird" (Everybody is co-responsible for the way he is governed). According to this view, the citizens of Stalinist Russia and fascist Germany were politically responsible for the actions of their leaders. They were co-responsible, along with the dictatorial parties, for their political life. It is not clear whether this shared responsibility derives from the unrealized ability to overthrow the dictator or follows simply from the fact that, as Jaspers put it, this is the political realm in which *ich mein Dasein habe* ("my existence is lived out").[91] On the latter theory, political guilt derives from identification with the society and one's presence in it at that time. Locke referred to this as tacit consent.

The argument for political guilt based on personal history resembles Freud's account of why we bear responsibility for the evil impulses of our dreams: "Unless the content of the dream...is inspired by alien spirits, it is part of my own being."[92] For the sake of effective therapy, we must accept our dreams as our own. The argument appeals to our desire for coherence and authenticity in our personality. The same demands of consistency require us to recognize that we are part of the culture that has nourished us. The theme of alienation runs throughout both Freud and Jaspers. We should not treat our dreams as alien to us, and we should not be aliens in our own land. This is the best way to understand Jaspers's claim that we accept co-responsibility for the way we are governed.

Jaspers thinks of metaphysical guilt as arising from solidarity with other human beings. The failure to rescue even with no prospect of success generates this form of existential guilt. As Jaspers describes the German situation, "We did not go into the streets when our Jewish friends were led away; we did not scream until we too were destroyed....We are guilty of being alive."[93] Metaphysical guilt goes beyond all other forms of guilt. As Jaspers argues, "Someplace between human beings there is room for the unconditional proposition that either we live together or we do not live at all."[94]

These propositions bring to mind a Talmudic analysis of sacrificing one to save many.[95] As the case is put, a Jewish caravan is surrounded by an enemy force. The enemy says, "Give us one of you as a hostage, or we kill everyone in the caravan." The rabbis conclude that the duty of the members of the caravan under these circumstances is to die together rather than arbitrarily to identify one of their number as a hostage.[96] Of course, this flies in the face of what many a utilitarian would decide. This example illustrates Jaspers's point that there are some situations in which the solidarity of human beings requires them to endure the same fate. Suppose the travelers remain passive as the enemy troops approach the caravan and arbitrarily pick a hostage. If they resist, they will all be killed. But failing to resist, failing to die, they become, as Jaspers claims, metaphysically guilty for the death of their compatriot.

But where does the duty of solidarity stop? Claims of metaphysical guilt are presumably limited to a particular cultural situation and therefore stop short of the universal guilt advocated by Father Zosima in *The Brothers Karamazov*.[97] If everyone is guilty for everything, then everyone is also innocent. The distinction loses its bite.

These arguments for political and metaphysical guilt offer a qualified defense of collective guilt. Jaspers defends the idea that some groups can be charged with political or metaphysical guilt, but he takes a strong stand against the idea that nations as such can bear guilt of this sort. He denies the guilt of the German people (as opposed to Hitler and his party) for the war and the Holocaust on the ground that the German nation has no clear contours. There is no way of knowing who is included and who is not, who is at the core of the nation and who is at the periphery. He does not deny that some people possess more or less of certain national characteristics, but nationality is a scalar—rather than categorical—concept. It is not like being male or female, but more like being tall or short. You can have more or less Germanness in your sense of identity, but there is no fixed level in this variable identification that defines someone as a German in his heart. There are many different Germans; no single identity can be reduced to a composite German. As he writes, "An entire nation [*Volk*] cannot be reduced to a single individual. A nation cannot suffer heroic tragedy. It cannot be a criminal; it cannot act morally or immorally. Only individuals in the nation can do these things."[98]

The clincher in Jaspers's rejection of German guilt is his reliance on anti-Semitism as the paradigmatic charge of collective guilt. It is irrational, he claims, to hold Jews liable in eternity because two thousand years ago a specific set of Jews in Jerusalem collaborated with the Romans in crucifying a man who later was called the Messiah. There is no doubt that anti-Semitism had its roots, in part, in centuries of calumny against Jews as Christ-killers. If we now understand this kind of undifferentiated indictment of a nation as irrational and bigoted, he argues, we should not repeat the mistake by charging all Germans with the crime of the Holocaust. Rhetorically and logically, Jaspers's point compels our attention. To ascribe irreducible, associative national guilt to the Germans is to repeat the intellectual indecency of anti-Semitism.[99]

We must also remember that collective guilt is reciprocal. It is not simply the loser or the rogue nation that suffers from the collective guilt associated with its actions. The complicated psychological structure that we have charted in this chapter is independent of one's status as the justified or unjustified party in a military conflict. What is done in the name of the citizen includes not just the launching of a war, but also one's conduct during the war—two logically independent elements of the law of war, as we have demonstrated in previous chapters. Why else would Americans argue so passionately about the extraordinary renditions and secret torture prisons operated by the CIA, or the invasion of Iraq? It is precisely because so much is at stake, not just for the nation but for ourselves as citizens, if we conclude that these actions are unjustified. It is not just the other who can suffer from collective guilt.

This examination of the collective dimension of guilt brings us full circle, back to our original discussion of the collective nature of crimes defined by the Rome Statute. The origins of collective guilt go back at least as far as the Bible, if not further, reemerged with vigor during the nineteenth century and its Romantic fascination with nationhood, and were central to German discussions about war guilt for the Holocaust. What emerges from these discussions is the central truth that our lives are embedded within a larger culture, and when our actions are performed in the name of that culture, they carry with them implications for our brethren. It is, of course, to overstate the point to call this "blood guilt," as it was during the Romantic era. We are not arguing for a return to the offensively racial or reductively ethnic notions that guided this period. But so, too, is it an exaggeration to so cling to the Enlightenment commitment to the individual that one forgets the cultural and national significance of one's actions. There is an undeniable connection between the individual and the collective when arms are taken up in defense of the state. It is therefore inevitable that, as we noted at the beginning of this chapter, the Rome Statute defines war crimes, crimes against humanity, and genocide as collective crimes. This is a more refined and sophisticated attempt to chart these connections, with the use of legal concepts, than the broad brushstrokes offered by the Romantic era. The project of international criminal law is, indeed, an attempt to bridge these very two aspects of warfare, the individual and the collective. The goal of international criminal justice is to apportion criminal blame to individuals for actions that are always pursued with the collective in mind, by loyalty to either a flag, a constitution, a political ideal, or, in more racist situations, a particular ethnicity.

Furthermore, this undeniable connection between the individual and the collective dimension of war explains why the analogy between individual and national self-defense has been so fruitful. The basic structure of legitimate defense is similar in the individual and collective realms for the very reason that both are tightly woven together in the very nature of warfare, where defense of self, defense of others, and defense of the nation all come together on the battlefield. What would be considered a purely private criminal action in peacetime becomes, in war, a collective action, triggering the machinery not just of criminal law but also of international law itself, for these crimes are truly of concern to the international community as a whole.

CONCLUSION

To some readers, our reliance on Romantic notions such as culture and nation will come as a surprise, to be sure, and at worst, a shock—perhaps even labeled dangerous. The reliance on such notions may be considered especially ironic in a book devoted to war and crime, violent endeavors fueled by the flames of national passion more than any other kindling. One might think, then, that we have relied on the very notions that we ought to have dispelled, that we conceded too much by operating within the very paradigm that is responsible for the cultural institutions of war and war crimes that we have studied. Why not reject such notions altogether, the skeptic might ask, and argue for a cosmopolitan ideal where all individuals are treated the same and national cultures evaporate from the scene, replaced by more individualistic and modern notions of humanity?

This argument has been made before, and usually it has been countered by the positive claim that something would be lost if national cultures disappeared and were replaced by one monolithic McWorld.[1] According-ing to one view, this would be disastrous for individuals, who by their very nature require a social context within which to flourish.[2] According-ing to a second view, national cultures have value in and of themselves, independent of the individuals who constitute them. Therefore, even if every member of a national culture were resituated in a new context, and were happy in it, this would still be a bad thing for humanity. The world would have lost something of value: a culture, a way of life, a set of values and attitudes that enrich the tapestry of human existence. The value of a national culture is not just instrumental, but intrinsic. National cultures are more than a source of warmongering and national prejudice, they are a source of the inspiration and difference that make human flourishing possible.

Although both of these responses to the skeptic have merit, we must set them aside for the moment and explain why we have lingered so long on the concepts of nationhood and guilt. In the preceding pages, we have attempted to offer a realistic account of human life, one that presents a balanced picture of what life is and what life might be. There are elements to human existence that are subject to revision—imagination is the only limit—and there are elements that are hardwired into our lives, whether by virtue of biology, language, or the structure of rational cognition. And there are elements that fall somewhere between, subject to revision in some possible way but highly unlikely to be revised because they are so deeply embedded in civilization.

National cultures, and the very concepts of nationhood and people, are exactly that, we have argued. We live our lives within these national cultures, and to suggest that the world could be replaced by a set of purely cosmopolitan connections is somewhat disingenuous, given the facts on the ground. In a sense, this is one manifestation of the well-known distinction in politics between ideal-world theory and real-world theory. In ideal-world theory, one can make optimistic assumptions about one's ability to rewrite the way things are; in real-world theory, one must remain more realistic and then train one's gaze on how to order the world in the face of these intractable realities. The existence of national cultures and peoples is one such notion. It is here to stay, with all of its benefits and its horrific missteps, and our task is to reorient our theory to offer the best possible outcome in light of these realities.

To that end, we have offered in these pages what amounts to an ironic turn. We have taken the concept of nationhood—a source of wars of aggression and crimes against humanity—and turned it into a concept that, when properly understood, provides the basic theoretical foundation for resisting these phenomena. The nation is recognized in our account as a basic element on which a claim of legitimate defense can be raised. A violent attack against a nation, whether or not it constitutes a state, is an illegal action that can be resisted, not only by the nation itself, but by the entire world community. Furthermore, we have shown in chapter 6 how this claim is embedded within international law and a proper reading of Article 51 and its recognition of a natural right to légitime défense.

Of course, it is also subject to great abuse, and we have taken great pains to define the outer limits of our own argument, explicitly demonstrating the logical flaws in any legal argument for preemptive war that jettisons the key concept of imminence. Although we have demonstrated how a war of humanitarian intervention might be justified by legitimate defense, it is always possible for a nation to marshal these legal resources as a ruse to justify its aggressive military actions. But this possibility should not foreclose the possibility of legitimate and honest appeals to these legal arguments. That any legal doctrine might be perverted by the powerful is no reason to deny its use by the weak.

There is a tendency to view the relationship between nations as an international state of nature, in the Hobbesian sense of the expression: nasty, brutish, and short. The conduct of nations is regulated by the world community, the United Nations, and our understanding of international law. But there is no Leviathan to enforce our edicts. The conduct of nation-states has always therefore been violent, and perhaps necessarily so, given the necessity for self-help actions in a situation where there is no government power to enforce the laws with the same immediacy as the domestic police. But we are reminded here of Cicero and his conclusion that "there is no more difference between this orderly civilized life, and that former savagery, than there is between law and violence."[3] So Cicero suggests that the state of nature was not some historical moment, but rather an ever-present and ongoing tension between respect for law and the resort to violence. The international law of self-defense, as we have outlined it in this book, stands on the border between these human forces. Referring again to the tension between law and violence, Cicero writes:

> If we prefer not to use one of these, we must use the other. Do we want violence eradicated? Then law must prevail, i.e., the verdicts in which all law is contained. Are the verdicts disliked or ignored? Then violence must prevail. Everyone understands this.

Of course, it would be preferable if the world community universally accepted the conclusions of international law. This is ideal-world theory if ever there was one. But there will always be rogue states, launching wars of aggression under the banner of self-defense, doing rhetorical violence to the concept in their quest to justify illegal actions. The right of legitimate defense, both on our own behalf and on the behalf of others, is real-world theory, an attempt to come to terms with the fact that one cannot always assume that the rest of the world will respect a nation's right to exist. Indeed, at this moment in time, it is clear that this respect is lacking. The current president of Iran has stated explicitly that Israel has no right to exist and should be wiped off the face of the map. These comments have spawned controversy among legal scholars who point out that President Ahmadinejad may be guilty of incitement to commit genocide under the Rome Statute.[4] Putting aside this controversy, however, one is left with the undeniable fact that when law fails, one must resort to violence. But this need not mean that in the wake of violence, law should fall by the wayside. As we have attempted to demonstrate in these pages, the use of legitimate defense is the lawful and justified use of violence, which can prevail in the face of a nation's worst nightmare: an attack from beyond its horizon, or even from within.

NOTES

INTRODUCTION

1. *See Rasul v. Bush*, 542 U.S. 466 (2004).

2. *Ex parte Quirin*, 317 U.S. 1 (1942).

3. This was Justice Antonin Scalia's conclusion in his dissenting opinion in *Hamdi v. Rumsfeld*, 542 U.S. 507, 569 (2004).

CHAPTER 1

1. Oscar Schachter, *Self-Defense and the Rule of Law*, 83 Am. J. Int'l L. 259 (1989).

2. The Nazis claimed self-defense on all fronts and asserted that the claim was beyond adjudication. This position was rejected at Nuremberg, thus establishing that the question of self-defense is always subject to legal analysis. *See* Judgment of the International Military Tribunal at Nuremberg, 1946, 1 *Trial of German Major War Criminals Before the International Military Tribunal* 208 (1947).

3. *See Dennis v. U.S.*, 384 U.S. 855 (1966).

4. Professor Wolfgang C. Friedmann, a sixty-five-year-old survivor of Nazi Germany, was stabbed during a mugging by three teenage assailants. *See* Emanuel Perlmutter, *Professor Slain in Mugging Here*, N.Y. Times, September 20, 1972, at A1.

5. *See generally*, George P. Fletcher, *A Crime of Self-Defense: Bernhard Goetz and the Law on Trial* (Chicago: University of Chicago Press, 1988).

6. David Cole, *Their Liberties, Our Security: Democracy and Double Standards*, 31 Int'l J. Legal Info. 290 (2003).

7. *See Military and Paramilitary Activities in and against Nicaragua*, 1986 ICJ Rep. 14, *reprinted in* 25 I.L.M. 1023 (1986); *Legal Consequences of the Construction of a Wall in the Occupied Palestinian Territory*, Advisory Opinion, 2004 ICJ Rep. 136.

8. *See* Treaty for the Renunciation of War as an Instrument of National Policy (Kellogg-Briand Pact), 94 L.N.T.S. 57 (August 27, 1928).

9. This analogy is explored in greater depth in chapter 6.

10. For the origin of this idea, see Emmerich de Vattel, *The Law of Nations*, edited by Joseph Chitty, § 12, at 3 (Philadelphia: T & J. W. Johnson, 1853) (1758); *see also* Immanuel Kant, *The Metaphysics of Morals*, edited by Mary Gregor, 114 (New York: Cambridge University Press 1996) (1797), describing the state "as a moral person...living in relation to another state in the condition of natural freedom."

11. *See* John Rawls, *A Theory of Justice* (Cambridge, Mass.: Harvard University Press, 1971).

12. Rawls famously referred to this as the veil of ignorance.

13. Michael Walzer, *Just and Unjust Wars* (New York: Basic Books, 1977) and Yoram Dinstein, *War, Aggression and Self-Defense* (New York: Cambridge University Press, 2001).

14. A full analysis of the doctrine is offered later in this chapter, *infra* note 49 and accompanying text.

15. Rome Statute Articles 8(2)(b)(i), (ii), (iii), and (iv).

16. *See* Thomas Nagel, *War and Massacre*, 1 Phil & Pub. Affairs 123 (1972), *reprinted in* Marshall Cohen, Thomas Nagel, and Thomas Scanlon, eds., *War and Moral Responsibility* 3, 11 (Princeton, N.J.: Princeton University Press, 1974) ("The law of double effect provides an approximation" to the relevant distinction, which is "between what we do to people and what happens to them as a result of what we do"); Walzer, *Just and Unjust Wars*, at 152 (interpreting the proper attitude requiring that "some degree of care be taken not to harm civilians").

17. There are some exceptions, notably in the work of Richard Wasserstrom. See his articles *The Relevance of Nuremberg*, 1 Phil. & Pub. Affairs 22 (1971) and *The Laws of War*, 56 Monist 1 (1972). *See also* R. B. Brandt, *Utilitarianism and the Rules of War*, 1 Phil & Pub. Affairs 145 (1972), *reprinted in* Cohen et al., *War and Moral Responsibility*, at 25 (discussing the rules of the Hague and Geneva Conventions as the choices a rational nation would make from behind a veil of ignorance).

18. For a very useful collection of articles, see Cohen et al., *War and Moral Responsibility*.

19. *See* Nagel, *War and Massacre*, at 14.

20. This Rawlsian formulation is similar to T. M. Scanlon's notion of "reasonable rejectability." *See* T. M. Scanlon, *What We Owe to Each Other* (Cambridge, Mass.: Harvard University Press, 1998).

21. One obvious fallacy of Nagel's approach is that it would permit civilians to target soldiers. Soldiers might understand why they are being shot at it, regardless of who is taking aim. For a discussion of this problem, see Walzer, *Just and Unjust Wars*, at 325–26.

22. For reflections on these modes of argument in the common law and civil law, see George P. Fletcher and Steve Sheppard, *American Law in a Global Perspective: The Basics* 33–35 (New York: Oxford University Press, 2005).

23. Nagel, *War and Massacre*, at 20–21.

24. Brandt, *Utilitarianism and the Rules of War*.

25. Rome Statute Art. 8(2)(b)(xvii) (poison) and 8(2)(b)(xi) (treacherous wounding or killing).

26. Kant, *Legal Theory* at 57.

27. Hague Convention (1907) Art. 29–31. Kant's ban on snipers might also seem a bit extreme in light of evolving practices of warfare. For a cinematic treatment of the theme, see the German film *Stalingrad* (dir. Sebastian Dehnhardt, 2003). Nothing in the film suggests that the role of the sniper was dishonorable.

28. As cited in Captain Joshua E. Kastenberg, USAF, *The Legal Regime for Protecting Cultural Property During Armed Conflict*, 42 A.F. L. Rev. 277, n.10 (1997).

29. The Law of War on Land, being Part III of the Manual of Military Law 42 (HMSO 1958), as quoted in Knut Dörmann, *Elements of War Crimes under the Rome Statute of the International Criminal Court: Sources and Commentary* (New York: Cambridge University Press, 2003), Art. 8(2)(b)(xvii) at 283.

30. Department of the Army, FM 27–10 The Law of Land Warfare, Point 37, at 18 (1956) (quoted in full).

31. Kant, *Legal Theory*, at 346–47 [Gregor at 152–54]; Robert Kolb, *Origin of the Twin Terms Jus ad Bellum/Jus in Bello*, Int'l Rev. of Red Cross 553 (1997).

32. Nagel, *War and Massacre*, at 3.

33. Nagel goes even further and muses about whether there is ever a justification for killing in warfare, when disabling the enemy might be a sufficient means of self-defense. Nagel, *War and Massacre*, at 21 n.11.

34. Jeff McMahan, *Innocence, Self-Defense, and Killing in War*, 2 J. Pol. Phil. 193, 196 (1994). Because McMahan does not believe that every "unjust combatant" is personally culpable, a more appropriate term would have been "wrongful combatant."

35. *See* Rawls, *A Theory of Justice*, at 98.

36. Rome Statute Art. 33.

37. Alan M. Dershowitz, *Reasonable Doubt* 37–38 (New York: Simon & Schuster, 1996).

38. Declaring that no quarter will be given means that surrendering soldiers will be shot rather than taken prisoner.

39. Kant, *Legal Theory* § 37, at 114. The term in German is *Rechtens*, which has no clear English translation. Translator Mary Gregor calls the opposite of "public Right" "Right in civil condition." *Id.*

40. *Id.*

41. *Id.* § 38, at 115.

42. The law refers to a buyer in this situation as a "bona fide purchaser for value."

43. Kant, *Legal Theory* § 39, at 118.

44. *Id.* § 40, at 119.

45. Actually, this is not entirely true in modern legal spheres. The common law will not enforce gift promises, and the civil law systems require these promises to be in writing and to be notarized. For analysis of the problem, see Fletcher and Sheppard, *American Law in a Global Perspective*, at 396–98.

46. *Lieber Code*, Article 21.

47. Nagel, *War and Massacre*, at 141 n.11.

48. McMahan, *Innocence, Self-Defense, and Killing in War*, at 197–98.

49. David Rodin, *War and Self-Defense* (New York: Oxford University Press, 2002), at 141–62.

50. This is how the terms individual and collective self-defense are used in Article 51 of the United Nations Charter.

51. Walzer takes a more nuanced approach when he analyzes the justification for shooting a soldier who is relaxing and smoking a cigarette. *See* Walzer, *Just and Unjust Wars*, at 141–43.

52. Protocol I of 1977 to the Geneva Conventions, Article 1(4) referring to common Article 2 of the Geneva Conventions, which says simply that the law of war applies whether the war is declared or not. The United States has not ratified this Protocol.

53. Walzer discusses this case under the heading of "war treason." Walzer, *Just and Unjust Wars*, at 177, 208. *See also* Lt. Cmdr. Patricia Zengel, *Assassination and the Law of Armed Conflict*, 134 Mil. L. Rev. 123, 133–36 (1991) (distinguishing the crime of war treason from treason proper).

54. Article 7 of the Rome Statute on Crimes against Humanity addresses attacks against the civilian population. Article 8 on war crimes is not clear about who is protected and who is not. Some grave breaches of the Geneva Convention are defined so that they might include crimes committed by civilians or groups of civilians against military personnel. A good example is

Rome Statute Art. 8 (2)(a)(vi) (willfully depriving a prisoner of war or other protected person of the rights of fair and regular trial).

55. Rome Statute Art. 28.

56. *Id.* Art. 27(1).

57. *See* George P. Fletcher, *Romantics at War: Glory and Guilt in the Age of Terrorism* (Princeton, N.J.: Princeton University Press, 2002).

58. *Lieber Code* Art. 21.

59. *Id.* Art. 22 ("The principle has been more and more acknowledged that the unarmed citizen is to be spared in person, property, and honor as much as the exigencies of war will admit").

60. Exodus 22:1.

61. Exodus 22:2.

62. Kant, *Legal Theory,* at 333 [Gregor at 142].

63. *Babylonian Talmud,* Tractate Sanhedrin 73a.

64. Thomas Aquinas, *Summa Theologiae* II-II, Q. 64, art. 7.

65. John Gardner, *Fletcher on Offenses and Defenses,* 39 Tulsa L. Rev. 817 (2004).

66. Rodin, *War and Self-Defense,* at 33; Wesley N. Hohfeld, *Some Fundamental Legal Conceptions as Applied in Judicial Reasoning,* 23 Yale L.J. 16 (1913); Wesley N. Hohfeld, *Fundamental Legal Conceptions as Applied in Judicial Reasoning,* 26 Yale L.J. 710 (1917).

67. This way of thinking is implicit in many codes. *See, e.g.,* MPC 2.04 ("necessary for the purpose of protecting himself against the use of unlawful force"); StGB § 34 ("in order to avert an imminent unlawful attack").

68. Kant, *Legal Theory,* at 230 [Gregor at 56]. *Cf.* Benjamin Cardozo's definition of due process as "ordered liberty" in *Palko v. Connecticut,* 302 U.S. 319, 325 (1937).

69. StGB § 34.

CHAPTER 2

1. *See* Michael S. Moore, *Placing Blame: A General Theory of the Criminal Law* 66 (New York: Oxford University Press, 1997); *see also* Kyron Huigens, *Fletcher's* Rethinking: *A Memoir,* 39 Tulsa L. Rev. 803, 811 (2004). In German this theory is called *Die Lehre von negativen Tatbestandsmerkmalen* ("theory of the negative elements of the definition"). *See* Claus Roxin, *Strafrecht: Allgemeiner Teil,* vol. 1, 231–323 (Munich: C. H. Beck, 3d ed. 1997).

2. Available online at http://www.dod.mil/news/May2003/d20030430milcominstno2.pdf.

3. Kyron Huigens refers to this element as "AJC." *See* Huigens, *Fletcher's Rethinking: A Memoir*, at 811.

4. Judgment of the Supreme Court in Criminal Cases, 61 RGSt 242 (1927).

5. Rome Statute § 31(1) refers back to § 21(1)(c), which permits judicial development of the law based on comparative analysis.

6. *See* Antonio Cassese, *International Criminal Law* 26 (New York: Oxford University Press, 2003) (arguing that customary law should be used to fill gaps left by treaty rules and international instruments).

7. Rome Statute, Art. 23.

8. H. L. A. Hart, *The Ascription of Responsibilities and Rights*, 49 Proc. Aristotelian Soc'y, 171, 180, 188–90 (1949).

9. John Gardner refers to this result as the "remainder" thesis. John Gardner, *Fletcher on Offenses and Defenses*, 39 Tulsa L. Rev. 817 (2004).

10. *Id.*

11. There is an important logical implication of this structure: Element six itself consists of several subelements. See the six required elements of legitimate defense discussed in chapter 4. If an affirmation of SD requires a finding of all six, then a negation of SD is logically satisfied if any one of the six is missing.

12. The term "sovereign" does not appear in the U.S. Constitution, and it appears in only one incidental reference in the proposed Constitutional Convention for Europe, regarding "the sovereign base areas of the United Kingdom of Great Britain and Northern Ireland in Cyprus." *See* Article VI-4(6)a ("Scope") of the final version of the draft Treaty Establishing a Constitution for Europe (CONV 850.03en), available online at http://european-convention.eu.int/docs/Treaty/cv00850.en03.pdf.

13. *See* Christine Gray, *International Law and the Use of Force* 30 (New York: Oxford University Press, 2d ed. 2004).

14. Corfu Channel Case, 1949 ICJ Rep. 4.

15. *Ploof v. Putnam*, 71 Atl. 188, 81 Vt. 471 (1908); German Civil Code § 904.

16. Gray, *International Law and the Use of Force*, at 43. The Belgian position of this issue parallels their aggressive approach to universal jurisdiction, which resulted in major controversies in 2002 and 2003. *See* George P. Fletcher, *Against Universal Jurisdiction*, 1 J. Int'l Crim. J. 579 (2003).

17. *Case Concerning Legality of Use of Force (Yugoslavia v. United States of America)*, 38 I.L.M. 1188 (June 2, 1999) (dismissing ICJ case for lack of jurisdiction).

18. Article 2 of each of the four Geneva Conventions includes the following statement: "The Convention shall also apply to all cases of partial or total occupation of the territory of a High Contracting Party, even if the said occupation meets with no armed resistance." In addition, the Fourth Geneva Convention contains an entire section (composed of Articles 47–78) on the

treatment of civilians in occupied territories. *Cf.* the provisions of the Hague Convention of 1907, Articles 42–56.

19. *See* Judith Jarvis Thomson, *The Realm of Rights* (Cambridge, Mass.: Harvard University Press, 1990), at 361–67.

20. Kant, *Legal Theory* at 348 [Gregor at 154].

21. As regards terrorism, the United Nations is currently operating under twelve different relevant conventions and protocols; see http://www.undcp .org/odccp/terrorism_definitions.html. The first official attempt at definition appears in the League of Nations Convention of 1937, which was never enacted: "All criminal acts directed against a State and intended or calculated to create a state of terror in the minds of particular persons or a group of persons or the general public." Terrorism expert A. P. Schmid proposed in 1992 that the UN simply adopt the stance that acts of terrorism are the "peacetime equivalents of war crimes." *Id.*

22. The common law of trespass conforms to the same categories as used in the law of necessary defense. *Trespass quare clausum fregit* covered trespass to land; *trespass vi et armis* applied to personal assaults; and *trespass de bonus asportatis* addressed the taking of chattels.

23. *See* John Fabian Witt, *Toward a New History of American Accident Law: Classical Tort Law and the Cooperative First-Party Insurance Movement*, 114 Harv. L. Rev. 690, 775 n.457 (2001).

24. John Donne: "No man is an island, entire of itself; every man is a piece of the continent, a part of the main." "Meditation XVII," from *Devotions upon Emergent Occasions* (1624).

25. The Thorns Case, 6 Edw. 4, f.7, pl. 18 (K.B. 1466).

26. *See* Victor E. Schwartz, Kathryn Kelly, and David F. Partlett, *Prosser, Wade, and Schwartz's Torts* 802–3 (New York: Foundation Press, 10th ed. 2000).

27. Admittedly, there are some cases in which the right of resistance must yield to the social interest in the intrusive activity. *See Boomer v. Atlantic Cement Co.*, 26 N.Y.2d 219 (1970).

28. Theodor Lenckner, Peter Cramer, Albin Eser, and Walter Stree, Schönke/Schröder, *Strafgesetzbuch Kommentar* § 32, points 5–9, at 604–6 (Munich: C. H. Beck, 26th ed. 2001) [hereafter cited as *Kommentar*].

29. Decision of the Bay, OLGSt, January 22, 1963, 1963 NJW 824.

30. The MPC recognizes justification for force in the execution of a public duty (§ 3.03), self-protection (§ 3.04), protection of other persons (§ 3.05), protection of property (§ 3.06), in law enforcement (§ 3.07), and by persons, "with Special Responsibility for Care, Discipline, or Safety of Others" (§ 3.08). In terms of self-protection, § 3.04(b) states, "The use of deadly force is not justifiable under this Section unless the actor believes that such force is

necessary to protect himself against death, serious bodily harm, kidnapping or sexual intercourse compelled by force or threat."

31. *See* George P. Fletcher, *Fairness and Utility in Tort Theory*, 85 Harv. L. Rev. 537, 564–69 (1972).

32. An entire volume of the ICJ Reports is devoted to the numerous opinions rendered in this case. *See* ICJ Reports (1986).

33. *See* the interpretation by Gray, *International Law and the Use of Force*, at 67; G.A. Resolution 2625.

34. This is collective self-defense in the sense used by international lawyers, not philosophers, as we explained in chapter 1.

35. MPC § 3.05.

36. *See Military and Paramilitary Activities in and against Nicaragua*, 1986 ICJ Rep. 14, *reprinted in* 25 I.L.M. 1023 (1986).

37. Incidentally, all of these words also refer to the right hand, and carry in addition the connotation of "making straight" or "rectifying."

38. The English word "justification" comes from the Latin *justificare*, meaning to justify. *See Oxford English Dictionary* at 328–29. The Latin *ficare*, or *facere*, means "to make," so that justification means literally "to make lawful."

39. *See* chapter 4, section 5.

40. StGB § 32.

41. Rome Statute § 31(1)(c).

42. Although the passage in Exodus 22:1 comes close to this: "If a thief is found breaking in, and he is struck so that he dies, there shall no blood be shed for him."

43. There is no authority for this proposition. For most people it is intuitively obvious. But if they disagree, it is unclear how to convince them, except to say that the incompatibility thesis is built into the rule of law.

44. The Convention does not explicitly protect nations, but rather "national, ethnical, racial or religious" groups against the intentional effort to destroy them "in whole or in part." This language is not incorporated in the Rome Statute Art. 6. *See also* ICCPR Art. 1 ("All peoples have the right of self-determination").

45. Kent Greenawalt, *The Perplexing Borders of Justification and Excuse*, 84 Colum. L. Rev. 1897, 1921 (1984).

46. The First Amendment provides that "Congress shall make no law . . . abridging the freedom of speech, or of the press." The principle of not censoring speech on the basis of content is expressed in a large body of case law. *See Cohen v. California*, 403 U.S. 15 (1971).

47. Kant, *Legal Theory*, Introduction, at § C.

48. Belgium argued this point before the ICJ. *See* Request for the Indication of Provisional Measures, Legality of Use of Force (*Yugoslavia v. Belgium et al.*), CR 99/14 (May 10, 1999). During the oral hearing, counsel for Belgium argued that "NATO, and the Kingdom of Belgium in particular, felt obliged to intervene to forestall an ongoing human catastrophe, acknowledged in Security Council resolutions. To safeguard what? To safeguard, Mr. President, essential values which also rank as *jus cogens*. Are the right to life, physical integrity, the prohibition of torture, are these not norms with the status of *jus cogens*?" *See id.* at CR 99/15.

49. H. Patrick Glenn, *Legal Traditions of the World: Sustainable Diversity in Law* 95 (New York: Oxford University Press, 2000).

50. A good example is David Weiss Halivni, *The Book and the Sword: A Life of Learning in the Shadow of Destruction* (New York: Farrar, Straus, & Giroux, 1996).

51. *Babylonian Talmud*, Tractate Sanhedrin 72a.

52. The law of rodef (aggressors) came into disrepute in 1995 when leading rabbis in Israel used the term to label Yitzhak Rabin as an aggressor against his own nation, by virtue of his plans to make peace with the Palestinians and evacuate territory seized in the Six-Day War. Yigal Amir reputedly acted on this politicized interpretation of Jewish law when he assassinated Rabin. Under Jewish law he was under a duty to kill the rodef. It is fair to say that this twisted version of the doctrine has no grounding in the Talmudic discussion. For one thing, the rabbis who declared Rabin a rodef were in effect declaring him to be an outlaw subject to assassination at any time; thus they themselves constituted aggressors relative to Rabin's life.

53. *Babylonian Talmud*, Tractate Sanhedrin 72a. This passage is unfortunately often invoked in political contexts—for example, to argue for militant actions against Palestinians—since the way the maxim is formulated and applied, there is no need to establish either imminence or necessity.

54. *Babylonian Talmud*, Avodah Zarah 26a, 26b.

55. This was the rule at common law. *See* George P. Fletcher, *Rethinking Criminal Law* 857 (New York: Oxford University Press, 1978, 2000).

56. *Babylonian Talmud*, Sanhedrin (last Mishna in chap. 4).

57. It is not surprising, then, that contemporary Orthodox Jewish ethicists have considerable difficulty with the distinction between active and passive euthanasia. *See* J. David Bleich, *Judaism and Healing: Halakhic Perspectives* 176–78 (Jersey City, N.J.: Ktav Publishing, 2002). Contrary to the growing consensus of the medical community, they deny that there is a morally significant difference between killing and letting die, between causing death and standing idly by as one's neighbor dies of natural causes. Every case of letting someone die violates the imperative of Leviticus 19:16. Every case of letting someone die sacrifices an entire world.

58. For a discussion of this, as well as two other cases of "permissive action" in the Talmudic framework, *see* Marilyn Finkelman, *Self-Defense and the Defense of Others in Jewish Law: The Rodef Defense*, 33 Wayne L. Rev. 1257 (1987), at 1284–86.

59. The Hebrew expression is *Lo Damim lo* (Exodus 22:1). This is sometimes translated, "There shall be no blood shed for him."

60. Robert M. Cover, *Obligation: A Jewish Jurisprudence of the Social Order*, 5 J. L. & Religion 65 (1987).

61. For approval of this practice, see Walzer, *Just and Unjust Wars*, at 207–11.

62. *See* International Law Commission, Draft Articles on Responsibility of States for Internationally Wrongful Acts, Art. 23, 24, 25, UN Doc. A/56/10 (2001) (precluding wrongfulness in cases of *force majeure*, distress, and necessity).

63. This classic American slogan originated on the revolutionary-era "rattlesnake" flags used by the various naval forces of the rebellious colonies, the most famous of which, known as the "Gadsden flag," sported a coiled rattlesnake on a yellow ground. The Williamsburg, Virginia, *Gazette* of May 11, 1776, commented on the Gadsden flag thus: "The colours of the AMERICAN FLEET to have a snake with thirteen rattles, the fourteenth budding, described in the attitude of going to strike, with this motto, "DON'T TREAD ON ME!" See *Trivia*, 33 Wm. & Mary Qtrly 528 (July 1976).

64. Sir Edward Coke, 3 *The Institutes of the Laws of England* 55 (1644).

65. 116 Cal. Rptr. 233, 526 P.2d 241, 12 C.3d 470 (1974). Some states have now passed controversial statutes to explicitly allow the use of deadly force to stop burglaries, even in the absence of a threat to serious bodily injury. *See, e.g.*, N.Y. Penal Law § 35.20(3).

CHAPTER 3

1. *Code Pénal* § 122–25. On the interpretation of légitime défense in Belgian law, see Françoise Tulkens and Michel van de Kerchove, *Introduction au Droit Pénal: Aspects Juridiques et Criminologiques* (Brussels: Story-Scientia, 5th ed. 1999).

2. *See, e.g.*, Carta de las Naciones Unidas, Art. 51 (translating self-defense as "*legitima defensa, individual o collectiva*"). The Russian version reveals some particularities. The term parallel to "inherent right" is *neot'emlemoe prava* or "inalienable right." The term for self-defense is *samooborona*, which is in fact a literal translation of "self-defense" rather than the term used in the Russian Penal Code, which is equivalent to the French concept.

3. *See* Rome Statute Art. 31(1)(c).

4. *See* Definition of Aggression Resolution, pmbl. G.A. Res. 3314 (1974) (reaffirming that "the territory of a State shall not be violated by being the object, even temporarily, of military occupation").

5. The point of referring to self-defense as an inherent right is that if the Security Council fails to act, states may still claim the right to act under Article 51. *See* Thomas M. Franck, *Recourse to Force: State Action against Threats and Armed Attacks* 48 (New York: Cambridge University Press, 2002). On the possible interpretation of Article 51 to include the automatic right to defend other states under attack, *see* D. W. Bowett, *Self-Defenses in International Law,* at 201 (New York: Praeger, 1958) (quoting French sources relying on the term légitime défense). Bowett recognizes the analogy to the domestic law of self-defense and in the end, on the basis of common law sources, rejects a universal right to defend others, even in the domestic law of self-defense. *Id.* at 202–3.

6. For a discussion of the legality of U.S. intervention in Vietnam under a theory of self-defense, see Louis Henkin, *How Nations Behave: Law and Foreign Policy* 306–8 (New York: Columbia University Press, 1979) (discussing whether the United States was coming to the defense of a victim of aggression or intervening in a civil war); John C. Bender, *Self-Defense and Cambodia: A Critical Appraisal*, 50 B.U.L. Rev. 130, 133–34 (1970) (discussing whether there was the necessary degree of coordination between the U.S., South Vietnamese, and Cambodian governments to establish a right of intervention based on collective self-defense).

7. This idea was suggested by Annemieke van Verseveld.

8. Indeed, overlapping alliances are often cited by historians as a contributing cause of World War I.

9. Article 5 of the treaty was invoked by the North Atlantic Council on September 12, 2001, the day after the attacks, and was reaffirmed on October 2, 2001, after an official investigation that determined that the attacks were conducted by Al Qaeda under the direction of Osama bin Laden. *See* North Atlantic Treaty, Art. 5 (April 4, 1949) ("The Parties agree that an armed attack against one or more of them in Europe or North America shall be considered an attack against them all and consequently they agree that, if such an armed attack occurs, each of them, in exercise of the right of individual or collective self-defence recognised by Article 51 of the Charter of the United Nations, will assist the Party or Parties so attacked by taking forthwith, individually and in concert with the other Parties, such action as it deems necessary, including the use of armed force, to restore and maintain the security of the North Atlantic area.").

10. Dinstein, *War, Aggression and Self-Defence*, at 225.

11. *Id.*

12. *Gulf of Tonkin Resolution*, 88th Congress, 2nd Sess., 78 Stat. 384 (Aug 10, 1964).

13. *See* J. William Fulbright, foreword to Michael J. Glennon, *Constitutional Diplomacy* (1990), discussed and cited in Philip Bobbitt, *Constitutional Lessons of Vietnam and Its Aftermath*, 92 Mich. L. Rev. 1364, 1365 (1994).

14. *See Military and Paramilitary Activities in and against Nicaragua*, 1986 ICJ Rep. 14, at 104, *reprinted in* 25 I.L.M. 1023 (1986).

15. *Id. See also* Morrison, *Legal Issues in the Nicaragua Opinion*, 81 Am. J. Int'l L. 160 (1987).

16. *See* Dinstein, *War, Aggression and Self-Defence,* at 238.

17. Latin American states had recently signed the Act of Chapultepec and were concerned about protecting it. *See* Ruth B. Russell, *A History of the United Nations Charter* (Washington, D.C.: Brookings Institution, 1958), at 698.

18. *See id.* at 698 (noting that U.S. officials were "concerned lest the new provision place too much emphasis on regionalism at the expense of the world organization").

19. *See id.* at 696.

20. *Id.* at 690.

21. *See* Covenant of the League of Nations, Art. 15(7) (1919).

22. *See* Russell, *A History of the United Nations Charter*, at 698.

23. *Id.* at 698.

24. *Id.* at 699.

25. *See* Leland M. Goodrich et al. (eds.), *Charter of the United Nations: Commentary and Documents* 354–55 (New York: Columbia University Press, 1969).

26. *See* Eric A. Feldman, *The Ritual of Rights in Japan* (New York: Cambridge University Press, 2000).

27. *See id.* at 16.

28. Feldman, *id.*, argues against the prevailing wisdom that Japanese legal culture places a premium on community and eschews claims based on individual rights.

29. *See* Russell, *A History of the United Nations Charter*, at 930.

30. Belgium argued that ultimate competence to interpret the Charter should rest with the General Assembly but that the ICJ could issue advisory opinions. Britain argued that each organ of the UN could provide its own interpretation of matters within its operational competence. *See* Russell, *A History of the United Nations Charter*, at 925–27.

31. The ICJ and the Security Council sometimes offer conflicting and mutually exclusive interpretations of the Charter. The court concluded in *Legal Consequences of the Construction of a Wall in the Occupied Palestinian Territory* that Article 51 did not apply in cases of military threats from nonstate actors. However, the Security Council has taken a different view and recognized the applicability of Article 51 in cases of terrorism. For an in-depth

discussion, see Kathleen Renee Cronin-Furman, note, *The International Court of Justice and the United Nations Security Council: Rethinking a Complicated Relationship*, 106 Colum. L. Rev. 435 (2005). *See also* Jose E. Alvarez, *Judging the Security Council*, 90 Am. J. Int'l L. 1, 6 (1996) (discussing judicial review by the ICJ of Security Council actions).

32. Rawls refers to this type of agreement as a mere modus vivendi, a Latin phrase meaning roughly "a way of living." This arrangement allows the parties to continue functioning together without resolving the underlying disagreements between them. *See* John Rawls, *Political Liberalism* 146 (New York: Columbia University Press, 1993).

33. *See* Franck, *Recourse to Force*, at 46.

34. *See id.* at 47 (noting that the proposal received more than 50 percent of the votes but that amendments required 66 percent support to pass).

35. *See* Jean Delivanis, *La Légitime Défense en Droit International Public Modern: Le Droit International face à ses limites* (Paris: Librairie Generale de Droit et de Jurisprudence, 1971).

36. *Id.* at 148.

37. This view is also expressed by H. Al Chalabi, *La Légitime Défense en Droit International* 101–9 (Cairo: Editions Universitaires D'Egypte, 1952).

38. *See* Delivanis, *Légitime Défense*, at 149 ("Cependant, il est nécessaire qu'il existe une relation relativement proche, tout au moins dans certains droits, entre les personnes qui se défendant en commun").

39. Code Pénal, Art. 122–25.

40. *Restatement (Second) of Torts*, Section 76.

41. *See* John Salmond, *Torts* 375 (London, Sweet & Maxwell, 11th ed. 1953), quoted in *Restatement (Second) of Torts*, Section 76.

42. *See Leward v. Basely* (1694) and *Tickel v. Read* (1772), *cited in* Bowett, *Self-Defenses in International Law*, at 201 n.3.

43. Indeed, Bowett seems to be the original source of this analysis, which is then repeated by Delivanis without further scrutiny or independent research.

44. Bowett, *Self-Defenses in International Law*, at 201–2. Bowett also cites Roman law for this proposition, including Sohm's *Institutes of Roman Law*. However, the fact that a proximate relationship was required in Roman law was based in the extreme inequities of Roman society. Masters could protect slaves because they were property, and men could protect their family because they had dominion over them. But the law has long since abandoned these master-slave dialectics. This point is also made by the late Josef Kunz, who concludes that "the right to defend others in municipal law is often restricted to persons having a special family relation with the person defended." *See* Josef Kunz, *Individual and Collective Self-Defense in Article 51 of*

the Charter of the United Nations, 41 Am. J. Int'l L. 872, 875 (1947). However, Kunz correctly notes that légitime défense includes defense of others and cites the French scholar Descamps for this proposition, as well as the similar distinction in German law between *Nothilfe* (defense of others) and the more general *Notwehr* (necessary defense). *Id.* Also, instead of recognizing an independent notion of defense of others under the umbrella of légitime défense, Kunz recognizes that collective self-defense could be a broad notion that might eventually allow the use of force for "any and all states members." *Id.* at 874.

45. Indeed, German law to this day still limits the excuse of necessity to those who stand in a close relationship with the defendant. *See* StGB Art. 35.

46. John Westlake, *Chapters on the Principles of International Law* 115 (Cambridge: Cambridge University Press, 1894) (concluding that the "act of self-preservation must in this as in all other cases be limited to what is strictly imposed by the emergency"). The French translations are instructive. Self-preservation was translated as *le droit international de conservation*. Westlake, *Traité de Droit International*, translated by A. de Laparadelle, 328 (Oxford: Oxford University Press, 1924) (but also referring to légitime défense within the context of protection on the high seas).

47. Most of the writers of this time period focus on self-preservation. Wheaton focuses on self-preservation, although he considers self-defense to be a more specific right within the larger category of self-preservation. *See* Henry Wheaton, *Elements of International Law* 90 (Boston: Little, Brown, 1866). The French translation for self-preservation is *le droit de conservation*, and the translation for self-defense is légitime défense. *See* Wheaton, *Éléments du Droit International* 75–76 (Leipzig: F. A. Brockhaus, 1874). Woolsey says that war can be waged to defend any right "analogous to the individual's right of self-preservation, and of defending his house when attacked." *See* Theodore D. Woolsey, *Introduction to the Study of International Law* 184 (New York: Scribner, Armstrong & Co., 5th ed. 1879).

48. *See* Westlake, *Chapters on the Principles of International Law*, at 111. But Westlake does defend a more limited right of self-preservation when a state "must take its defence into its own hands." *Id.* at 115.

49. The same concepts are expressed in Ellery C. Stowell, *Intervention in International Law* 392–405 (Washington, D.C.: John Byrne, 1921) (discussing self-preservation and necessity). Stowell alternates between describing necessity as a justification and as an excuse, although it is clear from his analysis that he rejects a broad defense of necessity in international law. He is also clear that it cannot excuse departure from rules of warfare "essential to prevent war from degenerating into an indiscriminate slaughter, like the contests of beasts." *Id.* at 404.

50. The case is discussed and quoted in Rodin, *War and Self-Defense*, at 111.

51. *See also* Stanimir Alexandrov, *Self-Defense against the Use of Force in International Law* 215–32 (The Hague: Kluwer Law International, 1996).

52. Bowett, *Self-Defenses in International Law*, at 202.

53. *See also* Hans Kelsen, *Collective Security and Collective Self-Defense under the Charter of the United Nations*, 42 Am. J. Int'l L. 783, 792 (1948) (noting that the terminology of collective self-defense is "problematical"). Kelsen too seems to restrict the right of collective self-defense to regional treaty arrangements.

54. *See* Ian Brownlie, *International Law and Use of Force by States* (Oxford: Clarendon Press, 1963) at 330–31.

55. *See id.* at 231, *citing* Verdross, 30 *Hague Recueil* (1929); Séfériadès, 34 *Hague Recueil* (1930); Cavaglierie, *Nuovi Studi sull' Intervento* (1928); Baty, *The Canons of International Law* (1930).

CHAPTER 4

1. *See* Fletcher, *Rethinking Criminal Law,* at 213–18, 223–26.

2. The Definition of Aggression Resolution was adopted by the General Assembly without a vote, and the U.S. government did not consider the resolution a binding codification of customary international law. *See* U.N. GA Res. 3314, UN Doc. A/RES/3314/Annexes (XXIX) (1974). For a discussion of the U.S. position regarding the resolution's legal status, see generally Sean D. Murphy, *Contemporary Practice of the United States Relating to International Law*, 95 Am. J. Int'l L. 387, 400–401 (2001) (discussing U.S. position on aggression during ICC negotiations).

3. MPC § 3.04.

4. This is the debate in the *Goetz* case, which was resolved in favor of the "objective" standard of reasonableness in *People v. Goetz*, 502 N.Y.S.2d 577 (N.Y. 1986).

5. Eric Schmitt, *Marine Set for Questioning in Death of Wounded Iraqi*, N.Y. Times, November 16, 2004, at A12.

6. Rome Statute § 31(1)(c).

7. *See* the analysis of the *Menendez* case in George P. Fletcher, *With Justice for Some: Victims' Rights in Criminal Trials* (Reading, Mass.: Addison Wesley, 1995), at 146–47.

8. This requirement of public evidence is discussed in greater detail in chapter 7.

9. MPC § 3.04; Rome Statute § 31(1)(c); StGB § 32.

10. For a defense of this analysis, see George P. Fletcher, *The Right and the Reasonable*, 98 Harv. L. Rev. 949 (1985).

11. The phrase "imminent and unlawful use of force" in 31(1)(c) appears in the French translation of the Rome Statute as *recours imminent et illicite à la force,* and in the Spanish as *un uso inminente e ilícito de la fuerza.*

12. MPC § 3.04(1).

13. Kant, *Legal Theory* at 347 [Gregor at 153].

14. *See* Brownlie, *International Law and Use of Force by States* 257–61, 366–67 (mixing up the terms "anticipatory action" and "preventive action" with the claim that the force must be directed toward an "imminent attack"). *See* the discussion of the *Caroline* incident below in the text.

15. Walzer, *Just and Unjust Wars,* at 81.

16. Kant, *Moral Theory,* at 4:400 [Gregor at 13].

17. *See* N.Y. Penal Law § 35.15(1).

18. For further discussion, see Fletcher, *A Crime of Self-Defense* at 29.

19. *Brown v. United States,* 256 U.S. 335, 343 (1921) (Holmes, J.).

20. For extensive analysis of both trials, see Fletcher, *With Justice for Some* at 37–68.

21. *See* interview on NPR's *Talk of the Nation* on May 2, 2005, concerning the new Florida law abolishing the duty to retreat. Available online at http://www.npr.org/templates/story/story.php?storyId=4627356.

22. MPC § 2.04(3)(b)(ii) ("actor knows that he can avoid the necessity of using such force with complete safety by retreating").

23. MPC § 2.94(3)(b)(ii)(1) (no duty to retreat from home or place of work).

24. *See* Mitchell Berman, *Justification and Excuse, Law and Morality,* 53 Duke L.J. 1, 13 (2003). In this otherwise thoughtful article, Berman ignores the requirement of necessity in self-defense cases and therefore claims, incorrectly, that the defense applies without moral justification, as, for example, in a case in which the defender could easily avoid the attack by stepping aside but does not.

25. Decision of September 20, 1920, 55 *RGSt.* 82 [hereafter cited as *Decision of September 20, 1920*].

26. Christine Gray, *International Law and the Use of Force* (Oxford: Oxford University Press, 2d ed. 2004), at 159.

27. *Cf.* Code Pénal § 122–25 (explicitly recognizing proportionality as a limitation).

28. This ethical duty is expressed in the principle *Rechtsmissbrauch* or "abuse of law." *See Kommentar* § 32, point 46, at 426.

29. For a detailed discussion of abuse of rights, see chapter 5, section 4.

30. ECHR, Art. 2(2)(a).

31. NYPL § 35.15.

32. This view is taken in Talmudic law, which reasons that the homeowner will resist the burglar, which in turn will escalate the conflict and lead to risk of deadly attack. *Babylonian Talmud,* Tractate Sanhedrin 72.

33. *See* the State of Colorado's "Make My Day" statute, Colorado Revised Statutes § 18-1-704.5 (2004).

34. For some doubts on the legitimacy of defending unoccupied territory, see Rodin, *War and Self-Defense*, at 134–37.

35. *See, e.g.,* Brownlie, *International Law and Use of Force by States*, at 261–64 (discussing proportionality as though it were synonymous with necessity).

36. Rome Statute § 8(2)(b)(4) (making it a war crime to "intentionally [launch] an attack in the knowledge that such attack will cause incidental loss of life or injury to civilians or damage to civilian objects or widespread, long-term and severe damage to the natural environment which would be clearly excessive in relation to the concrete and direct overall military advantage anticipated").

37. German Civil Code (BGB) § 228.

38. *Id.* There is an exception for cases in which the defender brings about the attack by, say, letting the dog out of confinement.

39. German Civil Code (BGB) § 904.

40. *Id.* The American law is typified by *Vincent v. Lake Erie Trans. Co.,* 109 Minn. 456, 124 N.W. 221 (1910).

41. For a difficult and wrenching case, see the problem of conjoined twins *In Re: A (children),* 4 All E.R. 961 (2000). The court decided in favor of separating the twins to save the one who had the substantially better chance of survival.

42. The same principle is reflected in the law of eminent domain. Objects that are a source of danger to others may be destroyed without paying compensation. There is an apparent analogy to the hypothetical of the rabid dog in passive necessity. *See United States v. Caltex* (Philippines), 344 U.S. 149 (1952).

43. *See* B. Sharon Byrd, *Putative Self-Defense and Rules of Imputation: In Defense of the Battered Woman,* 2 Jahrbuch für Recht und Ethik /Annual Review of Law and Ethics 283–306 (1994), reprinted in Leo Katz, Michael S. Moore, and Stephen J. Morse, eds., *Foundations of Criminal Law* (Oxford: Oxford University Press, 1999), at 260.

44. Notably Byrd, *id.*

45. *See* Glanville Williams, *Textbook of Criminal Law* 504 (London: Stevens, 2d ed. 1983); Paul H. Robinson, 2 *Criminal Law Defenses,* 12–29 (St. Paul, Minn.: West, 1984).

46. *See* Anne-Marie Slaughter, *Good Reasons for Going around the U.N.,* N.Y. Times, March 18, 2003, at A33.

47. Paul H. Robinson, *Competing Theories of Justification: Deeds vs. Reasons,* in A. T. H. Smith and A. Simester, eds., *Harm and Culpability* 45 (Oxford: Clarendon Press, 1996).

48. The more general problem is the theory of legal impossibility, which is addressed systematically in Fletcher, *Rethinking Criminal Law,* at 174–84. The essence of the critique is that the factor "unlawful invasion" is not relevant to the U.S. motivations, and therefore you cannot rationally attribute to the United States an effort to bring about an unlawful invasion.

CHAPTER 5

1. George P. Fletcher, *Proportionality and the Psychotic Aggressor: A Vignette in Comparative Criminal Theory,* 8 Israel L. Rev. 367 (1973).

2. *See* Robert Nozick, *Anarchy, State and Utopia* (New York: Basic Books, 1974), n.63. The article has not received as much attention from law professors, having been cited only about twenty times in the course of thirty years.

3. *See* Greenawalt, *Perplexing Borders;* George P. Fletcher, *The Psychotic Aggressor—A Generation Later,* 27 Israel L. Rev. 227 (1992).

4. Jerome Hall, *General Principles of Criminal Law* 436 n.85 (Indianapolis: Bobbs-Merrill, 2d ed. 1960); Glanville Williams, *Criminal Law: The General Part* 733 (London: Stevens, 2d ed. 1961). See also the parallel treatment in the French literature, Pierre Bouzat and Jean Pinatel, 1 *Traité de droit pénal et de criminologie* 363 (Paris: Dalloz, 1963); Henri Donnedieu de Vabres, *Traité de Droit Criminel et Législation Pénal Comparée* 232 (Paris: Recueil Sirey, 3d ed. 1947).

5. James Goldschmidt, *Der Notstand, ein Schuldproblem,* Oesterreichische Zeitsschrift für Strafrecht 129, 224 (1913).

6. *Regina v. Dudley and Stevens* (1884), 14 Q.B. 273 [hereafter cited as *Dudley & Stevens*].

7. StGB § 35. The conditions of the defense are that (1) there be an imminent risk of death or serious bodily harm, (2) the risk threatens the actor or his dependent, (3) the situation is not the actor's fault, and (4) there is no other way to avoid the risk but to inflict the harm in question.

8. The Court of Criminal Appeals in the United Kingdom recently upheld a killing justified as a choice of the lesser evil. *See In Re: A (children),* 4 All E.R. 961 (2000).

9. Most strikingly in Bouzat and Pinatel, *Traité de droit pénal et de criminology,* at 362; *compare* Hall, *General Principles of Criminal Law,* at 436 n.85 (analogy to natural force).

10. The same point was made by Alexander Loeffler, *Unrecht und Notwehr,* 21 Zeitschrift für die gesamte Strafrechtswissenschaft 537, 539 (1901) (criticizing those "who regard it as modern to depreciate the life of the insane").

11. Kant, *Legal Theory,* appendix to the introduction, § II.

12. For recent discussions of the relationship between culpability and character, *see* Douglas Husak, *Crimes outside the Core,* 39 Tulsa L. Rev. 755,

763–65 (2004); Kyron Huigens, *Virtue and Inculpation*, 108 Harv. L. Rev. 1423 (1995); Ekow Yankah, *Good Guys and Bad Guys: Punishing Character, Equality, and the Irrelevance of Moral Character to Criminal Punishment*, 25 Cardozo L. Rev. 1019 (2004).

13. *See* Meir Dan-Cohen, *Responsibility and the Boundaries of the Self*, 105 Harv. L. Rev. 959 (1991); Michael S. Moore, *Causation and the Excuses*, 73 Cal. L. Rev. 1091 (1985).

14. Coke, *The Institutes*, at 55; Matthew Hale, 1 *Pleas of the Crown* 481–87 (1680); Michael Foster, *Crown Law* 275 (1762); William Hawkins, 1 *A Treatise of the Pleas of the Crown* 113 (1716); 4 *Blackstone,* at 184.

15. Hawkins, *A Treatise of the Pleas of the Crown*, at 186; Foster, *Crown Law*, at 278; Coke, *The Institutes*, at 55.

16. Indeed, se defendendo was thought to be part of the theory of necessity. *See* 4 *Blackstone*, at 186.

17. It is generally held that the phrase first appeared in U. F. Berner, *Die Notwehrtheorie*, Archiv des Criminalrechts 547, 557, 562 (1848), though the maxim derives straight from the conceptual framework of Kant's theory of law. *See* George P. Fletcher, *Law and Morality: A Kantian Perspective*, 87 Colum. L. Rev. 533 (1987).

18. Coke, *The Institutes*, at 55.

19. John Locke, *Two Treatises on Civil Government* 120–27 (1690).

20. Tom Wolfe, *Bonfire of the Vanities* (New York: Bantam, 1988).

21. Locke, *Two Treatises*, at 126.

22. Rome Statute Art. 31.

23. On the structural difference between self-defense and necessity, compare Figures 4.1 and 4.2 supra.

24. For a good recent exposition of this theory, see Rodin, *War and Self-Defense*.

25. Some noteworthy German writers maintain that self-defense is based on a theory of balancing interests. Theodor Lenckner, *Gebotensein und Erforderlichkeit der Notwehr*, 1968 GA I.

26. 4 *Blackstone* at 182. This seems to be the first account of why there are limits to the common law privilege of deadly force.

27. There is a certain paradox to this interpretation of the common law. The system has been notoriously slow in recognizing necessity as a justification, particularly in England. *See, e.g., Dudley & Stephens* (rejecting necessity as a defense for shipwrecked sailors who resorted to cannibalism). Necessity was only recently recognized as a defense in, for example, *Re A (children)*, where British doctors separated conjoined infants so that the viable twin might live. For a general discussion of developments in the common law defense of necessity, *see* George P. Fletcher, *Basic Concepts of Criminal Law* (New York: Oxford University Press, 1998), at 130–32.

28. *Decision of September 20, 1920.*

29. This way of formulating the Kantian idea of Right or Law is analogous to John Rawls's First Principle of Justice: "Each person is to have an equal right to the most extensive basic liberty compatible with a similar liberty for others." John Rawls, *A Theory of Justice* 60 (Cambridge, Mass.: Belknap Press of Harvard University, 1971).

30. *Kommentar* § 32, n.44, at 601.

31. Hale, *Pleas of the Crown*, at 485–86 (distinguishing between trespass and felonies); Calif. Penal Code § 197 (1) (permitting deadly force "when resisting any attempt to commit a felony"). This sweeping language has been narrowed in the case law, *e.g.*, *People v. Jones*, 191 Cal. App. 2d 478, 481; 12 Cal. Rptr. 777 (1961) (privilege limited to felonies involving danger of serious bodily harm). *See* 4 *Blackstone*, at 180 (deadly force limited to "forcible and atrocious crimes"); *Storey v. State*, 71 Ala. 329 (1882) (limited to "atrocious crime[s] committed by force").

32. Friedrich Oetker, *Notwehr und Notstand*, in August Hegler, ed., *Festgabe für Reinhard von Frank zum 70* (Tübingen: J. C. B. Mohr, 1930), at 359.

33. Eberhard Schmidhäuser, *Strafrecht: Allgemeiner Teil, Lehrbuch* (Tübingen: Mohr, 1970), at 196 (in the author's words, these attacks do not put "the empirical validity of the Legal Order in question").

34. For an early discussion of the development of the abus de droit doctrine in private law, *see* H. C. Gutteridge, *Abuse of Rights*, 5 Cambridge L.J. 22, 24–29 (1935) (tracing the historical development of the term in German and French private law).

35. *Brown v. United States*, 256 U.S. 335, 343 (1921) (Holmes, J.).

36. *See* Decision of the Bavarian Higher State Court. Coercion (or *Noetigung*) is punished under StGB § 240.

37. GG § 103 (II).

38. One reply to this argument is that imposing limits on the defense only makes explicit the underlying theory of the defense; *see* Lenckner, *Gebotensein und Erforderlichkeit der Notwehr*, at 8–9. This argument assumes, without support, that the German theory of self-defense reflects the balancing of interests.

39. *See, e.g.*, Case of Albrecht, Kessler and Streletz, 2 BvR 1851/94 et al. (individual responsibility and liability for killing fugitives at the inner-German frontier) (October 24, 1996), translated and reprinted in 18 H. R. L.J. 65, 69 (concluding that § 27 of the Border Act could be displaced by higher law "if it expressed a manifestly gross breach of basic ideas of justice and humanity").

40. For a discussion of the traditional handling of the defense of property interests in German law, *see* Fletcher, *The Right and the Reasonable*, at 967–70

(explaining the Kantian basis for the traditional German rule that permits resistance against even minor aggression). Again, this rule follows the German maxim "Right need never yield to Wrong."

41. Rome Statute Art. 31(1)(d).

42. *Id.* Art. 31(1)(c).

43. *Id.*

44. *Compare* MPC § 2.09 ("coerced…the use of, or a threat to use, unlawful force against his person or the person of another").

45. *Id.*; StGB § 35.

46. Rome Statute Art. 31(1)(d).

47. *Erdemović*, Case No. ICTY-96–22, Sentencing Judgement of the Trial Chamber II, March 5, 1998.

48. *Erdemović*, Case No. ICTY-96–22, Appeals Chamber Judgement, Dissenting Opinion of Judge Cassese, October 7, 1997. Cassese argued, "After finding that *no specific international rule* has evolved on the question of whether duress affords a complete defence to the killing of innocent persons, the majority should have drawn the only conclusion imposed by law and logic, namely that the *general* rule on duress should apply—subject, of course, to the necessary requirements. In logic, if no exception to a general rule be proved, then the general rule prevails." However, the majority made the opposite inference. Having found no customary rule allowing duress as a complete justification for killing innocent civilians, the court declined to permit the defense. Cassese does, nonetheless, concede that it would be difficult for a duress argument to succeed in cases where the victims are innocent civilians, because the defense applies only when the crime is not disproportionate to the evil sought to be avoided. Our analysis helps explain why: proportionality means different things in different legal contexts. See earlier discussion. The analysis of disproportionality in the context of duress depends, ultimately, on the theory for recognizing the claim—whether it is treated as a justification or an excuse.

49. Rome Statute Art. 31(3), taken together with Art. 21(1)(c).

50. For some reason, philosophers have a difficult time grasping this truth. For a detailed discussion, see chapter 1, section 4.

51. *See* Neil A. Lewis, *Justice Memos Explained How to Skip Prisoner Rights*, N.Y. Times, May 21, 2004, at A10. For a discussion of the role that lawyerly arguments played in justifying the unjustifiable, see Christopher Kutz, *Lawyers Know Sin: Complicity in Torture*, in Karen J. Greenberg, ed., *The Torture Debate in America* (New York: Cambridge University Press, 2005).

52. This is strictly prohibited as a crime against humanity under Rome Statute Art. 7(1)(i).

53. *See* Rome Statute Art. 7, and chapter 8 below.

54. These are strictly prohibited as crimes against humanity under the Rome Statute Art. 7 (1)(g) and as war crimes under Art. 8(2)(b)(xxii) (international armed conflicts) and Art. 8(2)(3)(vi) (subnational armed conflicts).

55. The issue will be addressed in volume 3 of George P. Fletcher, *The Grammar of Criminal Law: American, Comparative, and International* (Vol. 1, New York: Oxford University Press, 2007).

56. Rome Statute Art. 21(1)(c) (recognized under the "general principles of law derived from national laws of legal systems of the world").

57. *See* Greenawalt, *Perplexing Borders; see also* Francisco Munoz Conde, *"Rethinking" the Universal Structure of Criminal Law*, 39 Tulsa L. Rev. 941, 952 (2004). For the full argument that putative self-defense should be classified as an excuse, *see* Fletcher, *The Right and the Reasonable*.

58. A few scholars have begun to look at the issue of excusing nations based on reasonable mistakes of fact. *See* chapter 7.

CHAPTER 6

1. For a memoir of the peacekeeping mission, see Roméo Dallaire, *Shake Hands with the Devil* 146–47 (New York: Vintage Books, 2003). Dallaire writes, "My failure to persuade New York...still haunts me."

2. *See* Philip Gourevitch, *The Genocide Fax*, The New Yorker (May 1998).

3. The Security Council authorized military intervention in Somalia with S.C. Res. 794 on December 3, 1992. The Chapter VII authorization was based on the "deterioration of the humanitarian situation in Somalia" and the need for a "secure environment for humanitarian relief operations."

4. For a discussion of an obligation to intervene, see Gray, *International Law and the Use of Force*, at 40.

5. For example, the UN report on Darfur commissioned by the secretary general concluded that the war crimes and crimes against humanity committed in the Darfur region probably did not meet the legal requirements of genocide because the attacks were not motivated by genocidal intent. See *Report of the International Commission of Inquiry on Darfur to the United Nations Secretary-General* 130–31 (2005).

6. For a critical discussion of the legality of the Kosovo intervention, see Louis Henkin, *Kosovo and the Law of "Humanitarian Intervention,"* 93 Am. J. Int' L. 824 (1999) ("In my view, unilateral intervention, even for what the intervening state deems to be important humanitarian ends, is and should remain unlawful"). *See also* Gray, *International Law and the Use of Force*, at 38.

7. For a discussion of the relationship between moral and legal arguments for humanitarian intervention, see Fernando R. Teson, *Humanitarian Intervention: An Inquiry Into Law and Morality* (Dobbs Ferry, N.Y.: Transnational Publishers, 1988).

8. For a detailed consideration of jury nullification, see chapter 9 in Fletcher, *A Crime of Self-Defense*.

9. *See, e.g.*, Antonio Cassese, Ex iniuria ius oritur: *Are we Moving towards International Legitimation of Forcible Humanitarian Countermeasures in the World Community?* 10 Eur. J. Int'l L. 1 (1999). Cassese argues that intervention can be justified only if it is in response to a serious crime against humanity.

10. For a discussion of this problem, see Antonio Cassese, *Return to Westphalia? Considerations on the Gradual Erosion of the Charter System*, in *The Current Legal Regulation of the Use of Force* 515–20, edited by Antonio Cassese (Leiden: Brill Academic Publishers, 1986). Cassese argues that the erosion of UN control has gradually allowed states to reappropriate rights that they had under the Westphalian system and eventually lost when the UN system was created in 1945.

11. Cassese makes an argument for humanitarian intervention, but the argument is external to the UN Charter. *See also* Bartram S. Brown, *Humanitarian Intervention at a Crossroads*, 41 Wm. & Mary L. Rev. 1683, 1740 (2000) (referring to the legal indeterminacy of humanitarian intervention); Daphne Richemond, *Normativity in International Law: The Case of Unilateral Humanitarian Intervention*, 6 Yale Hum. Rts. & Dev. L.J. 45 (2003) (arguing that unilateral humanitarian intervention is surrounded by a "normative ambiguity" about its permissiveness under international law, not a strict legal regime governing its use).

12. *Legal Consequences of the Construction of a Wall in the Occupied Palestinian Territory*, Advisory Opinion, 2004 ICJ Rep. 136.

13. For a discussion of the relationship between self-determination and military intervention, see W. Michael Reisman, *Coercion and Self-Determination: Construing Charter Article 2(4)*, 78 Am. J. Int'l L. 642 (1984) ("Though all interventions are lamentable, the fact is that some may serve, in terms of aggregate consequences, to increase the probability of the free choice of peoples about their government and political structure"). A similar view is expressed by Anthony D'Amato in *The Invasion of Panama Was a Lawful Response to Tyranny*, 84 Am. J. Int'l L. 516 (1984) (arguing against a "statist" conception of international law in favor of one that recognizes "the reality of human beings struggling to achieve basic freedoms"). Reisman's view is sharply criticized in Oscar Schacter, *The Legality of Pro-democratic Invasion*, 78 Am. J. Int'l L. 645 (1984).

14. For an exhaustive analysis of customary rules regarding self-determination, see Antonio Cassese, *Self-Determination of Peoples* 67–99 (Cambridge: Cambridge University Press, 1995).

15. For a good discussion of self-determination as a legal right, see Heather A. Wilson, *International Law and the Use of Force by National Liberation Movements* 55–88 (Oxford: Clarendon Press, 1988). *See also* Ian Brownlie, *An Essay in the History of the Principle of Self-determination*, in *Grotian Society Papers: Studies in the History of the Law of Nations*, edited by C. H. Alexandrowicz (The Hague: Martin Nijhoff, 1968); Cassese, *Self-Determination of Peoples*.

16. *See* Gregory H. Fox, *The Right to Political Participation in International Law*, 17 Yale J. Int'l L. 539 (1992).

17. For example, the 1974 *Definition of Aggression* states that nothing in the Definition should "prejudice the right of self-determination, freedom and independence, as derived from the Charter, of peoples forcibly deprived of that right ... particularly peoples under colonial and racist regimes or other forms of alien domination; nor the right of these peoples to struggle to that end and to seek and receive support, in accordance with the principles of the Charter."

18. For a general discussion of the theories of nationalism, *compare* Ernest Gellner, *Nationalism* (New York: New York University Press, 1997) *with* Elie Kedourie, *Nationalism* (London: Hutchinson, 1960). On the role of nationalism in the American Civil War, see George P. Fletcher, *Our Secret Constitution: How Lincoln Redefined American Democracy* (New York: Oxford University Press, 2001). *See also* John Fabian Witt, *Patriots and Cosmopolitans: Hidden Histories of American Law* (Cambridge, Mass.: Harvard University Press, 2007).

19. Kedourie puts the point this way: "The doctrine [of nationalism] divides humanity into separate and distinct nations, claims that such nations must constitute sovereign states, and asserts that the members of a nation reach freedom and fulfillment by cultivating the peculiar identity of their own nation and by sinking their own persons in the greater whole of the nation." *See* Kedourie, *Nationalism*, at 71–72.

20. *But see* Christopher Heath Wellman, *A Theory of Secession* (New York: Cambridge University Press, 2005) (defending expansive theory of self-determination that would allow almost any unit to secede).

21. *See generally* Allen E. Buchanan, *Secession: The Morality of Political Divorce from Fort Sumter to Lithuania and Quebec* (Boulder, Colo.: Westview Press, 1991).

22. *See* John Locke, *Second Treatise of Government* 77, chap. XIII (Indianapolis: Hackett, 1980).

23. For a discussion of the concept of the nation as a cluster, see David Laitin, *Identity in Formation: The Russian-Speaking Populations in the Near*

Abroad (Ithaca, N.Y.: Cornell University Press, 1998). For a discussion of cluster concepts in general and their role in legal arguments, see Jens David Ohlin, *Is the Concept of the Person Necessary for Human Rights?* 105 Colum. L. Rev. 209, 229–31 (2005).

24. Philosophers sometimes express this as an inability to define a concept *per genus and differentiam.*

25. *See* Ludwig Wittgenstein, *Philosophical Investigations,* translated by G. E. M. Anscombe (New York: Macmillan, 1958).

26. For an extensive discussion of moral reactive attitudes, see P. F. Strawson, *Freedom and Resentment, in Free Will,* edited by Gary Watson (New York: Oxford University Press, 1982). Strawson's essay is a contribution to the literature on free will and determinism, and he argues that there is a limit to which we can rework our moral reactive attitudes to take into account a new theory. Therefore, even if we decided that humans have no free will and are not responsible for their actions, we would be unable to stop feeling gratitude or resentment toward others.

27. For a discussion of the contingent nature of nations and nationalism, see Benedict Anderson, *Imagined Communities* (London: Verso, 1991). Although nations are arbitrary constructs, it still may be the case that they play an important part in our lives. *See also* Eric Hobsbawm, *The Nation as Invented Tradition,* in *Nationalism* 76–83 (New York: Oxford University Press, 1994). For a stronger argument that nations and nationalism "remain the only realistic basis for a free society of states in the modern world," see Anthony D. Smith, *Nations and Nationalism in a Global Era* (Oxford: Polity Press, 1995).

28. Plato advanced his political theory and metaphysics by relying on an analogy between the divisions of the soul and the divisions of the city. *See* Plato, *The Republic,* translated by G. M. A. Grube (Indianapolis: Hackett, 1992).

29. Although individuals often exhibit significant rational disunity, it is nonetheless true that they demonstrate a commitment to rational unity, in the sense that most humans strive to unify their thoughts into a coherent mental picture, even if they ultimately fail to achieve it. Carol Rovane argues that this commitment to overall rational unity is what unifies a person over time. See her book, *The Bounds of Agency* (Princeton, N.J.: Princeton University Press, 1998), for a full treatment of this thesis.

30. The Enlightenment idea of the human being as a rational agent, with beliefs, desires, and values that can be coherently harmonized, is extremely important to economics, rational choice theory, and the law and economics movement. For an example of a political theory based on rational choice theory, see David Gauthier, *Morals by Agreement* (Oxford: Clarendon Press, 1986) (reworking traditional social contract theory with basic principles of rational choice).

31. *See* John Rawls, *The Law of Peoples* (Cambridge, Mass.: Harvard University Press, 1999). It should be noted that Rawls argued for a rule of nonintervention, though he argues that it should be "qualified" in cases of serious human rights abuses. Indeed, it should be clear from our analysis that our defense of humanitarian intervention is entirely consistent with the structure of *The Law of Peoples*. All peoples would consent to a social contract that allows intervention when their very existence is threatened, either through genocide or crimes against humanity.

32. *See id.* at 23–24.

33. J. S. Mill, *Considerations*, chap. XVI, *cited in* Rawls, *The Law of Peoples*, at 23 n.17.

34. Consider the following statement in 1987 of the Danish delegate to the UN General Assembly: "Wherever the exercise of the right to self-determination is violated, it is only natural that the matter be dealt with in the world organization. The denial of this right anywhere is a concern of peoples anywhere." UN Doc. 87/349. This statement, which offers a defense of military intervention in support of self-determination, is quoted in Cassese, *Self-Determination of Peoples*, at 156–57, although Cassese argues that military intervention, based on the denial of self-determination, is grounded in practical considerations, not legal assessment.

35. *See The Responsibility to Protect: Report of the International Commission on Intervention and State Sovereignty* (2001). The commission was created by the government of Canada, and the report was presented to the United Nations.

36. *See id.* at § 6.30 ("Although the General Assembly lacks the power to direct that action be taken, a decision by the General Assembly in favour of action, if supported by an overwhelming majority of member states, would provide a high degree of legitimacy for an intervention which subsequently took place, and encourage the Security Council to rethink its position").

37. The problem brings to mind the quote attributed to Charles de Gaulle: "How can anyone govern a nation that has 246 different kinds of cheese?"

38. The Belgian trusteeship of Rwanda had UN approval.

39. For a description of this process, see Philip Gourevitch, *We Wish to Inform You That Tomorrow We Will Be Killed with Our Families* 55–56 (New York: Farrar, Straus, & Giroux, 1998).

40. *See Developments in the Law—International Criminal Law*, 114 Harv. L. Rev. 1943, 2020 (2001).

41. Convention on the Prevention and Punishment of the Crime of Genocide, Art. 2 (1948).

42. The originator of the term genocide is, of course, Raphael Lemkin, who identified the nation and its culture as the interests protected by the

prosecution of "genos-cide" as a distinctive crime. *See* Raphael Lemkin, *Axis Rule in Occupied Europe* (Washington, D.C.: Carnegie Endowment for International Peace, 1944). The moral gravity of the crime stemmed not just from the murder of large numbers of people (which would be the same as crimes against humanity, after all), but the resulting loss to the world from cultural eradication.

43. The issue is discussed in W. D. Verwey, *Humanitarian Intervention*, in Cassese, *Current Legal Regulation of the Use of Force*, at 46.

44. *Compare* Christopher Hitchens, *A Long Short War: The Postponed Liberation of Iraq* (New York: Plume, 2003) (championing intervention in Iraq to stop human rights abuses committed by Hussein's regime), *with* David Rieff, *At the Point of a Gun: Democratic Dreams and Armed Intervention* (New York: Simon & Schuster, 2006) (expressing disillusionment with humanitarian justifications for military intervention in Iraq).

CHAPTER 7

1. For an early example, see Grotius, *On the Law of War and Peace (De jure belli ac pacis)*, book II, chap. 26. Reisman argues that the modern idea of preemption predates Bush and originates with the Shultz doctrine, formulated under the Reagan administration. *See* W. Michael Reisman and Andrea Armstrong, *The Past and the Future of the Claim of Preemptive Self-Defense*, 100 Am. J. Int'l L. 525, 528 (2006).

2. Immanuel Kant, *The Metaphysics of Morals*, § 54, in *Kant: Political Writings*, translated by H. B. Nisbet, 165 (Cambridge: Cambridge University Press, 2d ed. 1991).

3. *Id.* § 56, at 167.

4. See Immanuel Kant, *Perpetual Peace: A Philosophical Sketch*, in Kant, *Kant: Political Writings,* at 98.

5. Kant, *Metaphysics of Morals*, § 60, at 170.

6. *Id.*

7. Kant argues that we should always act in conformance with a maxim that could logically become a law for all mankind. This is the famous categorical imperative. See generally his *Grounding for the Metaphysics of Morals,* translated by James W. Ellington (Indianapolis: Hackett, 1981).

8. Dinstein, *War, Aggression, and Self-Defense*, at 172.

9. *See Transcript of the Candidates' First Debate in the Presidential Campaign*, N.Y. Times, October 1, 2004.

10. Surprisingly, international lawyers do not defend the blockade under principles of international self-defense. Dean Acheson rejected the relevance

of international law to Kennedy's decision to impose the blockade. *See* Franck, *Recourse to Force*, at 99–101. Schachter, *Self-Defense and Rule of Law*, at 260. Quincy Wright, *The Cuban Quarantine*, 57 Am. J. Int'l L. 546, 555–56 (1963) (concluding that the U.S. blockade was illegal because, inter alia, it violated Charter obligation to resolve disputes through peaceful means).

11. *See* Robert F. Kennedy, *Thirteen Days* (New York: W. W. Norton, 1971).

12. *See* Carl von Clausewitz, *On War* (New York: Penguin Classics, 1982).

13. *See* George W. Bush, Commencement Address at the United States Military Academy at West Point (June 1, 2002).

14. MPC § 3.04. Note that this section does not in itself impose a requirement of reasonable belief. The requirement becomes clear, however, by reading § 3.04 alongside the exceptions provided in § 3.09.

15. *See* N.Y. Penal Code § 35.15; 18 Pa. Cons. Stat. § 505; 720 Ill. Comp. Stat. 5/7–1; Mont. Code Ann. § 45–3-102; Oregon. Rev. Stat. § 161.209; Texas Penal Code § 9.32.

16. For a discussion of the syndrome, see Martha R. Mahoney, *Legal Images of Battered Women: Redefining the Issue of Separation*, 90 Mich. L. Rev. 1 (1991).

17. *See* Fletcher, *With Justice for Some*, at 133–37.

18. For examples, see generally Alafair S. Burke, *Rational Actors, Self-Defense, and Duress: Making Sense, Not Syndromes out of the Battered Woman*, 81 N.C. L. Rev. 211 (2002); Anne M. Coughlin, *Excusing Women*, 82 Cal. L. Rev. 1 (1994).

19. *But see* Cathryn Jo Rosen, *The Excuse of Self-Defense: Correcting a Historical Accident on Behalf of Battered Women Who Kill*, 36 Am. Univ. L. Rev. 11 (1986) (arguing that battered woman syndrome should be classified as an excuse because as a category excuses are less narrowly applied than justifications).

20. The idea has been discussed in the scholarly literature. *See* Kimberly Kessler Ferzan, *Defending Imminence: From Battered Women to Iraq*, 46 Ariz. L. Rev. 213, 215 (2004); Shana Wallace, Comment, *Beyond Imminence: Evolving International Law and Battered Women's Right to Self-Defense*, 71 U. Chi. L. Rev. 1749 (2004); John Yoo, *Using Force*, 71 U. Chi. L. Rev. 729, 753 (2004) (discussing the attempted redefinition of imminence in cases of battered wives); Michael Skopets, Comment, *Battered Nation Syndrome: Relaxing the Imminence Requirement of Self-Defense in International Law*, 55 Am. Univ. L. Rev. 753, 756 (2006); Jane Campbell Moriarty, *"While Dangers Gather": The Bush Preemption Doctrine, Battered Women, Imminence, and Anticipatory Self-Defense*, 30 N.Y.U. Rev. L. & Soc. Change 1 (2005).

21. *See* Security Council Resolution 487 (June 19, 1981) (strongly condemning the attack and calling it a violation of the UN Charter).

22. However, Paul Wolfowitz stately publicly that the Bush administration settled on the WMD justification out of "bureaucratic convenience" because it was the one argument that everyone could agree on. *See* Robert Jensen and Rahul Majahan, *End the Deception*, USA Today, June 4, 2003, at A10.

23. *See* Ferzan, *Defending Imminence: From Battered Women to Iraq*, at 228.

24. For a critical argument against the concept of rogue states, see Jacques Derrida, *Rogues: Two Essays on Reason*, translated by Pascal-Anne Brault and Michael Naas (Stanford: Stanford University Press, 2005). Mixed in with Derrida's animated and radical critique of U.S. foreign policy is an illuminating deconstruction of our use of the term "rogue." Although he makes the strong claim that the United States is, in fact, the real rogue state par excellence, this political claim can be divorced from the conceptual claim that the idea of the rogue state is an outlaw designation used by the strongest to vindicate their military actions.

25. *See* Kant, *The Metaphysics of Morals*, § 60, at 118–19. Kant defines an unjust enemy as a nation whose actions do not follow the categorical imperative. Specifically, unjust nations act in accordance with maxims that, when universalized, make peace impossible and throw the community of nations back into a perpetual state of nature.

26. *See, e.g.*, Lon Fuller, *The Morality of Law* (New Haven: Yale University Press, 1969) (arguing that laws must be, inter alia, general, public, and consistent).

27. Carl Schmitt, *The Concept of the Political,* translated by George Schwab (Chicago: University of Chicago Press, 1996).

28. This distinction is explored most thoroughly in the work of the German scholar Günter Jakobs. *See* his *Strafrecht allgemeiner Teil: die Grundlagenund die Zurechnungslehre* (Berlin: De Gruyter, 1983).

29. This notion goes as far back as Vattel, if not further. See Emmerich de Vattel, *The Law of Nations,* edited by Joseph Chitty, § 12, at 3 (Philadelphia: T & J. W. Johnson, 1853) (1758).

30. *See, e.g.*, D'Amato, *The Invasion of Panama Was a Lawful Response to Tyranny.*

31. In a 2002 memo to White House Counsel Alberto Gonzales, Deputy Assistant Attorney General John Yoo argued that federal law prohibited only torture that was "primarily" intended to inflict severe physical or mental pain or suffering, implying that torture primarily for the purposes of interrogation against Al Qaeda or Taliban detainees (having the incidental effect of causing pain and suffering) would not violate federal law. *See* John Yoo, *Memo to White House Counsel* (August 1, 2002). Furthermore, Yoo argued that the administration's war on terror would not be constrained by the international Torture Convention because the United States attached to its ratification an

"understanding" that imported this definition of torture from federal law. In a similar Justice Department memorandum, Assistant Attorney General Jay Bybee argued that federal law limited torture to cases of intense pain similar to that experienced during "serious physical injury, such as organ failure, impairment of bodily function, or even death." Mental suffering would constitute torture under federal law only if it produced "significant psychological harm of significant duration, e.g., lasting for months or even years." Regardless of this interpretation, though, Bybee argued that the executive branch was completely unrestrained by the federal law because it "impermissibly encroached on the President's constitutional power to conduct a military campaign." Bybee even cited Article 51 of the UN Charter and its inherent right of self-defense as another possible justification for the practice. *See* Jay Bybee, *Memorandum for Alberto R. Gonzales Re: Standards of Conduct for Interrogation under 18 U.S.C. §§ 2340–2340A* (August 1, 2002), n.26, at 45–46. For a discussion of these memos, see Jeremy Waldron, *Torture and Positive Law: Jurisprudence for the White House*, 105 Colum. L. Rev. 1681 (2005).

32. There is a lengthy literature on the morality and legality of so-called ticking bomb scenarios, where an imminent attack can be averted only by torturing the suspect for information. For a dubious defense of torture that fails to adequately respect the principle of reciprocity, see John Yoo, *War by Other Means: An Insider's Account of the War on Terror* (New York: Atlantic Monthly Press, 2006). For a more sophisticated discussion of torture, see Florian Jessberger, *Bad Torture—Good Torture?* 3 J. Int'l Crim. Just. 1059 (2005) (discussing a German case of police officers torturing a suspect to find kidnapping victim); see also *Public Committee Against Torture in Israel v. Israel*, Judgment Concerning the Legality of the General Security Service's Interrogation Methods, 38 I.L.M. 1471 (Sept. 6, 1999).

33. *See* Fletcher, *The Right and the Reasonable*, at 967. *Cf.* Kent Greenawalt, *The Perplexing Borders of Justification and Excuse*, 84 Colum. L. Rev. 1897, 1909 (1984) (arguing that many mistakes should be regarded as full justifications, not just excuses).

34. *See, e.g.*, Ferzan, *Defending Imminence: From Battered Women to Iraq*, at 220.

35. *See* Thom Shanker, *On Tour with Rumsfeld, the Jacket Stays on and the Monkeys Stay Away*, N.Y. Times, June 16, 2002, at A6.

CHAPTER 8

1. Benedict Anderson, *Imagined Communities: Reflections on the Origin and Spread of Nationalism* (London: Verso, 2d ed. 1991).

2. *Id.* at 39–40.

3. See Bates's assertion in *Henry V*: "We know enough if we know we are the king's subjects. If his cause be wrong, our obedience to the king wipes the crime of it out of us." William Shakespeare, *Henry V*, Act IV, Scene 1, in *The Oxford Shakespeare* (Oxford: Oxford University Press, 1914).

4. *Lieber Code*, Art. 20.

5. Fourth Hague Convention (1907), Art. 23(a) (poison) and 23(b) (treachery).

6. A similar incident is discussed in Marcel Ophuls's film, *The Sorrow and the Pity* (TV Rencontre, 1971).

7. Walzer expresses some sympathy for French farmers who killed German soldiers in the incident described in *The Sorrow and the Pity*. Walzer, *Just and Unjust Wars*, at 176–79. There is some support for this view. *See* Hilaire McCoubrey, *International Humanitarian Law: The Regulation of Armed Conflicts* 135 (Aldershot, England: Dartmouth, 1990) ("International law does not in itself prohibit the commission by inhabitants of an occupied territory of acts hostile to the belligerent occupant"). Palestinians regularly make this argument in defending "acts hostile" to the Israeli occupying force.

8. Rome Statute Art. 8(1) ("when committed as a part of a plan or policy or as part of a large-scale commission of such crimes"). This is described as a "grave breach under all the Geneva Conventions." William J. Fenrick, *War Crimes*, in Otto Triffterer, ed., *Commentary on the Rome Statute of the International Criminal Court* 173, 182 (Baden-Baden: Nomos, 1999), and if it met the other conditions it would violate Art. 8(2)(a)(i) of the Rome Statute.

9. *See Decision by the High Court of Justice Regarding the Assassinations Policy of the State of Israel* (December 14, 2006).

10. The relevant provision here is Article 51(3) of the Additional Protocol Relating to the Protection of Victims of International Armed Conflicts (1977), Protocol I to the Geneva Conventions of 1949 ("Civilians shall enjoy the protection afforded by this section, unless and for such time as they take a direct part in hostilities").

11. James Bennet, *Leader of Hamas Killed by Missile in Israeli Strike*, N.Y. Times, March 22, 2004, at A1 (describing the assassination of Sheik Ahmed Yassin); Greg Myre, *Leader of Hamas Killed by Israel in Missile Attack*, N.Y. Times, April 18, 2004, at A1 (reporting the assassination of Yassin's successor, Dr. Abdel Aziz Rantisi).

12. See *Ex parte Quirin*, 317 U.S. 1 (1942).

13. Some recent cases that use the term to describe the person being detained include *Padilla v. Bush*, 233 F. Supp. 2d 564 (2002), at 592; *U.S. v. Lindh*, 212 F. Supp. 2d 541(2002), at 554; *Gherebi v. Bush*, 374 F. 3d 727 (2003), at 729; *Hamdi v. Rumsfeld*, 124 S. Ct. 2633 (2004), at 2640; *Khalid v. Bush*, 355 F. Supp. 2d 311 (2005), at 319.

14. This is called the principle of complementarity under Rome Statute Art. 17(1).

15. Rome Statute Art. 12(2)(a), Art. 12(3).

16. *Id.* § 8(2)(a)(i).

17. *See, e.g.,* Statut de Rome de la Cour pénale internationale, Art. 8 (translating "wilful killing" as *l'homicide intentionnel*); Estatuto de Roma de la Corte Penal Internacional, Art. 8 (translating "wilful killing" as *el homicidio intencional*).

18. Rome Statute § 8(1).

19. *Id.* § 5, § 17(1)(d).

20. *D.T. v. Ind. Sch. Dist.,* 894 F.2d 1176, 1186 (10th Cir. 1990).

21. We have received guidance on this issue from Professor Antonio Cassese of Florence, Italy, formerly president of the International Criminal Tribunal for the Former Yugoslavia, in particular during conversations in early July 2001 in Paris. Professor Cassese is passionately committed to the view that all homicides committed by members of the occupying force constitute war crimes.

22. Rome Statute Art. 8(2)(b)(xxi).

23. *See* Fourth Geneva Convention Art. 6 ("The Occupying Power shall be bound, for the duration of the occupation, to the extent that such Power exercises the functions of government in such territory, by the provisions of [specifically mentioned] Articles of the present Convention").

24. There is no reason to classify the crime as a war crime just because we might be skeptical of justice in an Iraqi court under U.S. occupation. A court-martial proceeding against the American officer would also be possible. *See* 10 U.S.C. 817. Furthermore, failure of the United States to adequately supervise or punish a large number of soldiers guilty of wrongdoing might change the analysis and invoke issues of command responsibility.

25. Fourth Hague Convention (1907), Art. 3; Fourth Geneva Convention Art. 146.

26. 18 U.S.C. 2441.

27. Hague IV (1907), Art. 23(a), 23(b), 23(e).

28. *Id.* Art. 3.

29. *Id.* The compensation is presumably owed to the belligerent party of which the victim was a national.

30. Lieber refers to various crimes against domestic law but to only one crime against the law of nations—namely, taking and selling slaves from conquered troops. *See Lieber Code,* Art. 58. The Hague Convention of 1907 does not use the word "crime" at all.

31. On the relationship between aggression, unlawful force, and the "inherent right of self-defense," see chapter 3.

32. *See* U.N. GA Res. 3314 (defining aggression).

33. *See R v. Jones*, 2006 UKHL 16. The appellants in the case were arrested for demonstrating against the Iraq war and argued that the war's illegality under international law provided them with an affirmative defense against the charge of trespassing. Although the House of Lords agreed that aggression was a violation of customary international law, they concluded that the international prohibition against aggression was not incorporated into British domestic law and that the appellants could not appeal to an international crime to ground their criminal law defense.

34. The House of Lords noted that aggression was established as a violation of the international order as far back as the Pan-American Conference of 1928 and the Kellogg-Briand Pact.

35. Rome Statute Art. 8(2)(a)(vi). The major commentaries seem to be indifferent to this problem.

36. *Id*. Art. 8(2)(b)(vi).

37. The use of force is implicit in the other half of Article 8(2)(b)(viii), which prohibits "deportation or transfer of all or parts of the population of the occupied territory within or outside this territory."

38. *But see Legal Consequences of the Construction of a Wall in the Occupied Palestinian Territory*, Advisory Opinion, 2004 ICJ Rep. 136.

39. This principle is recognized in the Rome Statute Art. 22. Note that Art. 22(2) provides that "in case of ambiguity, the definition shall be interpreted in favor of the person being investigated, prosecuted or convicted."

40. Rome Statute Art. 7(1).

41. *Id*. Art. 7(2)(a).

42. The problem is similar to the relationship between the individual states and American citizens before the enactment of the Fourteenth Amendment in 1868. Prior to the Civil War, states could abuse their own citizens (reduce them to a state of slavery, etc.) provided they did not deny to citizens of other states the "privileges and immunities" accorded to the citizens of their own. U.S. Const. Art. IV, Sec. 2., Cl. 1. The Fourteenth Amendment shifted the focus from the protection of citizens to the protection of persons. U.S. Const., Amend. XIV ("nor shall any State deprive any person of life, liberty, or Property, without due Process of law; nor deny to any person within its jurisdiction the equal Protection of the laws").

43. The term, referring to "the killing of a people," was coined by Rafael Lemkin in 1944.

44. The Belgian experience is instructive. In early June 2001, a Belgian court convicted four Rwandans, including two Catholic Hutu nuns, of complicity in the murder of seven thousand Tutsis seeking refuge in their monastery. The *Economist* noted that "this was the first time that a jury of

citizens from one country had judged defendants for war crimes committed in another." *See Judging Genocide,* Economist, June 16, 2001, special section. The jurisdiction of the Belgian court was based on Article 7 of the Law of June 16, 1993, as amended on February 10, 1999, granting Belgian courts jurisdiction over genocide, as defined in the Genocide Convention, regardless of where the offense occurred, or by whom, or against whom the offense was committed. *See also* Beth Van Schaack, *In Defense of Civil Redress: The Domestic Enforcement of Civil Rights Norms in the Context of the Proposed Hague Judgments Convention,* 42 Harv. Int'l L.J. 141, 145 (2001).

45. The term "eliminationist" came into currency with Daniel Goldhagen's influential best-seller, *Hitler's Willing Executioners: Ordinary Germans and the Holocaust* (New York: Knopf, 1996).

46. This point was made in the Harvard note, *Developments in the Law—International Criminal Law,* 114 Harv. L. Rev. 1943, 2020 (2001).

47. *See Lieber Code* § 20.

48. For an extensive analysis of these Romantic impulses, see Fletcher, *Romantics at War,* as well as chapter 6 in this volume on humanitarian intervention (defending the Romantic conception of nationhood).

49. *See* Walzer, *Just and Unjust Wars,* at 172–75.

50. *See Case Concerning the Application of the Convention on the Prevention and Punishment of the Crime of Genocide (Bosnia and Herzegovina v. Serbia and Montenegro),* 2007 ICJ 91 (February 26, 2007).

51. This customary rule is also reflected in Article 4 of the International Law Commission's *Articles on State Responsibility.*

52. The Genocide Convention requires signatories (including Serbia) to take all necessary steps to prevent genocide and to punish (or extradite) all individuals responsible for genocide.

53. *See Bosnia v. Serbia,* § 388 ("The Court notes first that no evidence has been presented that either General Mladić or any of the other officers whose affairs were handled by the 30th Personnel Centre were, according to the internal law of the Respondent, officers of the army of the Respondent—a *de jure* organ of the Respondent. Nor has it been conclusively established that General Mladić was one of those officers; and even on the basis that he might have been, the Court does not consider that he would, for that reason alone, have to be treated as an organ of the FRY (Federal Republic of Yugoslavia) for the purposes of the application of the rules of State responsibility").

54. *See Bosnia v. Serbia,* § 387 ("The Applicant has shown that the promotion of Mladić to the rank of Colonel General on 24 June 1994 was handled in Belgrade, but the Respondent emphasizes that this was merely a verification for administrative purposes of a promotion decided by the authorities of the Republika Srpska").

55. *See Dissenting Opinion of Vice-President Al-Khasawneh*, § 3. The vice president also criticized the court for failing to demand access to Serbia's records from the Supreme Defense Council, which might have further established a direct link to the Serbian government.

56. *See* Amended Indictment, *Prosecutor v. Milosevic*, Case No. IT-02–54-T (April 21, 2004), § 32 (alleging a campaign of genocide against Bosnian Muslims).

57. Dershowitz, *Reasonable Doubt,* at 177. For a more recent rendering of Dershowitz's position, see his *Preemption: A Knife That Cuts Both Ways* (New York: W. W. Norton, 2007).

58. The Jews attribute a war crime, perhaps the first recorded crime, to the tribe of Amalek. They attacked the Jews from the rear when the Jews were "famished and weary." Deuteronomy 25:18. Nathan Lewin, *Deterring Suicide Killers*, Sh'ma: A Journal of Jewish Responsibility, May 2002, at 11–12.

59. This is the Fall River rape case, made famous by the film *The Accused.*

60. Dershowitz, *Reasonable Doubt,* at 176.

61. *See, e.g.*, 18 U.S.C. § 2 (2005) (punishing as a principal any individual who "aids, abets, counsels, commands, induces or procures" offenses against the United States).

62. Dershowitz, *Reasonable Doubt,* at 177.

63. *Id.* at 179.

64. Kant, *Legal Theory*, § 57.

65. Exodus 17:16: "The Lord will have war with Amalek from generation to generation." The discussion of a struggle between the descendants of Amalek and the people called Jews (rather than Hebrews) begins in the book of Esther, written in the early eighth century BCE.

66. A possible exception is Numbers 24:20: Amalek is described as a first among the nations but with a fate of "everlasting perdition."

67. We are indebted to Dr. Zvi Blanchard for this point.

68. Some translations of the Bible translate *avon* as "punishment," others as "sin" or "crime." The problem is well summarized in *Etz Chayim* [Tree of Life]: *Torah and Commentary*, edited by David Lieber, 27 n.13 (Philadelphia: Jewish Publication Society of America, 2001) (comment on the editor's choice of the word "punishment").

69. Sophocles, *Oedipus Rex*, edited by Stanley Appelbaum, translated by George Young (Mineola, N.Y.: Dover 1991). The text is not clear whether the pollution derives primarily from the patricide or the incest. The following lines of the Chorus suggest that the incest is at least a major factor: "Time found thee out—Time who sees everything—Unwittingly guilty; and arraigns thee now consort ill-sorted, unto whom are bred sons of thy getting, in thine own birthbed, O scion of Laius's race." *Id.* at 43.

70. Bernard Williams, *Shame and Necessity* 88–89 (Berkeley: University of California Press, 1993).

71. Aristotle, 3 *The Nicomachean Ethics,* translated by D. Ross, 1110–11b2 (London: Oxford University Press, 1925).

72. For the conventional view, see Paul Ricoeur, *The Symbolism of Evil,* translated by Emerson Buchanan, 100–102 (Boston: Beacon Press, 1969).

73. As the astute German commentator Claus Westermann points out, this was a startling new political idea—namely, that one brother could acquire a superior political status over his siblings. *See* Claus Westermann, *Die Joseph-Erzählung: Elf Bibelarbeiten zu Genesis* 24, 37–50 (Minneapolis: Augsburg Fortress Publishers, 1990).

74. *Compare* Reuben's later speech in Genesis 42:22 ("Did I not tell you, 'Do no wrong to the boy'? But you paid no heed. Now we must give an accounting for his blood").

75. With his usual political insight, Westermann points out that spying is a characteristic feature of nations, not of families. *See* Westermann, *Die Joseph-Erzählung,* at 73. In the first translation of the Bible into German, Martin Luther opted for a different term altogether. He translated the Hebrew term as *Kundschafter,* which means something like "investigator." *Die Heilige Schrift,* 1 Mose 42:9 (Gideon ed. 1967).

76. The Westermann thesis explains this response, but it seems strained nonetheless. Individuals and informal groups do, in fact, spy on each other. A totally different interpretation of the accusation might begin with the motive that Joseph attributes to the spying, namely, "to see the nakedness of the land." The French Jewish translator Andre Chouraqui captures the sexual dimensions nicely in his translation: "Vous etes venus pour voir le sexe de la terre." *La Sainte Bible,* entete 42:12 (Paris: Desclée de Brouwer, 1989). The sexual association is missing in Luther's translation, where the passage is rendered as the "investigators" coming to see "where the land is open." *Die Heilige Schrift,* at 42:12. The sexual overtones of the word "nakedness" suggest an analogy with the earlier intervention of the brothers—the "sons of Jacob"—to reclaim their sister Dinah from the house of Shechem. Whether or not that rescue was deceitful and improper, the brothers demonstrated their loyalty to members of their clan. By suggesting sexual overtones to the mission of his ten brothers, Joseph might be betraying his own yearning for them to have come to take him home as they had schemed and fought to hold onto Dinah. Rabbi David Silber suggested this interpretation.

77. *See* Genesis 42:21 ("And they said to one another, But we are guilty [*asham,* in the adjectival plural form *ashemim*] concerning our brother").

78. Some English translations unfortunately translate *asham* as "punished" simply because, as we noted, when guilt is understood as pollution, the decontaminating sacrifice is also called *asham* or guilt. *Etz Chayim,* at 260 n.21.

79. *See* Harry G. Frankfurt, *Freedom of the Will and the Concept of the Person*, in *The Importance of What We Care About* 20 (New York: Cambridge University Press, 1988). Frankfurt argues that the hallmark of personhood is the ability to form second-order beliefs and desires. For example, an addict who desires drugs may also have a desire (a wish) that he not have the desire to use drugs at all. This desire is *second-order* because it is a desire about another desire. Furthermore, although the first-level desire may be, in some way, compulsory, the second-level desire is entirely a function of our *attitudes* about our compulsions and therefore expresses our freedom of will.

80. *Lieber Code*, § 20.

81. The right of self-determination is explicitly recognized in Article 1 of the International Covenant on Civil and Political Rights ("All peoples have the right of self-determination. By virtue of that right they freely determine their political status and freely pursue their economic, social and cultural development"). For an excellent discussion of the concept of self-determination, see generally Avishai Margalit and Joseph Raz, *National Self-Determination*, 87 J. Phil. 439 (1990).

82. William Wordsworth, *Lyrical Ballads*, at xiv (Woodstock, N.Y.: Woodstock Books, 1997) (1800). A conventional dating of the movement relies on the publication of the first edition of Wordsworth's *Lyrical Ballads* in 1798.

83. William Wordsworth, *The Convention of Cintra*, in *The Prose Works of William Wordsworth*, edited by W. J. B. Owen and Jane Worthington Smyser, 191 (New York: Oxford University Press, 1974) (1809).

84. *See* Isaiah Berlin, *Three Critics of the Enlightenment: Vico, Hamann, Herder* 272 (Princeton, N.J.: Princeton University Press, 2000).

85. Frederick Charles von Savigny, *Of the Vocation of Our Age for Legislation and Jurisprudence*, translated by Abraham Hayard (Delran, N.J.: Legal Classics Library, 1986) (1831).

86. The original German version, *Was Is der Mensch*, was recently republished in translation as Karl Jaspers, *The Question of German Guilt*, translated by E. B. Ashton (New York: Fordham University Press, 2000).

87. Walter Russell Mead, *In the Long Run: Keynes and the Legacy of British Liberalism*, Foreign Affairs, January–February 2002, at 199, 203 (referring to "the notorious Morgenthau Plan, which would have reduced postwar Germany to a pastoral economy").

88. Whether punishment of the guilty is always necessary is one of the issues raised by the Joseph story, discussed in section 5 of this chapter.

89. *Dudley & Stephens*; Robinson, *Competing Theories of Justification: Deeds vs. Reasons*, at 125; Joshua Dressler, *Exegesis on the Law of Duress: Justifying the Excuse and Searching for Its Proper Limits*, 62 S. Cal. L. Rev. 1331, 1374–75 (1989).

90. *Dudley & Stephens*, at 288 n.17.

91. Jaspers, *The Question of German Guilt*, at 31. The translation in the text is our own. Ashton prefers "under whose order I live."

92. Sigmund Freud, *Some Additional Notes on Dream Interpretation as a Whole* (1925), reprinted in 19 *The Standard Edition of the Complete Psychological Works of Sigmund Freud*, edited and translated by James Strachey, 127, 133 (New York: W. W. Norton, 1961).

93. Jaspers, *The Question of German Guilt*, at 72.

94. *Id.* at 32. The translation is again our own. Ashton writes, "Somewhere among men the unconditioned prevails—the capacity to live only together or not at all."

95. *Jerusalem Talmud*, Tractate Terumot 8:4. According to the analysis in this passage, if the aggressor names a particular suspect and threatens to kill the travelers if he is not turned over, it is permissible to surrender him to save the caravan. The assumption is that if the authorities name a suspect, they have reasonable grounds to believe that he is guilty of some wrongdoing. Also, if the suspect is named, the guilt of choosing the victim does not fall upon the caravan.

96. *See generally* David Daube, *Collaboration with Tyranny in Rabbinic Law* (London: Oxford University Press, 1965).

97. Fyodor Dostoevsky, *The Brothers Karamazov*, translated by Richard Pevear and Larissa Volokhonsky (New York: North Point Press, 1990) (1880).

98. Jaspers, *The Question of German Guilt*, at 41. The translation in the text is our own. Ashton prefers: "One cannot make an individual out of a people. A people cannot perish heroically, cannot be criminal, cannot act morally or immorally; only its individuals can do so."

99. It might be possible to sustain a view of collective guilt that avoids this unfortunate conclusion. Such a view would require "non-transitive associative guilt"; that is, nations are guilty for their past actions but this fact alone entails nothing about the guilt of today's citizens. *See* George P. Fletcher, *The Storrs Lectures: Liberals and Romantics at War: The Problem of Collective Guilt*, 111 Yale L.J. 1499, 1549 (2002).

CONCLUSION

1. The term originates with Benjamin Barber, *Jihad vs. McWorld* (New York: Crown, 1995).

2. This view is exemplified in the work of Charles Taylor in *Sources of the Self: The Making of Modern Identity* (Cambridge, Mass.: Harvard University Press, 1992) and the communitarianism of Michael Sandel in *Liberalism and*

the Limits of Justice (New York: Cambridge University Press, 1998), although it is also defended from within the confines of political liberalism by some theorists, notably Will Kymlicka in *Liberalism, Community and Culture* (Oxford: Clarendon Press, 1989).

3. Cicero, *Pro Sext.* 1. 42. This quote, and its relationship to the state of nature, is discussed in David Hume, *An Enquiry Concerning the Principle of Morals* (Indianapolis: Hackett, 1983), at 24–25 n.11.

4. The discussion raises a serious jurisprudential question. Can an individual be prosecuted for incitement to commit genocide under the Rome Statute, even before a genocidal campaign was launched? The International Criminal Tribunal for Rwanda prosecuted individuals for incitement to commit genocide in Rwanda, but in that case the incitement caused an actual attack. The Iranian situation poses the question of whether the charge of incitement to commit genocide, under the Rome Statute, could be used as a tool for early intervention in a developing situation of concern to the international community.

INDEX